SENSUOUS GEOGRAPHIES

SENSUOUS GEOGRAPHIES

Body, sense and place

Paul Rodaway

London and New York

First published 1994
by Routledge
11 New Fetter Lane, London EC4P 4EE

Simultaneously published in the USA and Canada
by Routledge
29 West 35th Street, New York, NY 10001

© 1994 Paul Rodaway

Typeset in Garamond by Solidus (Bristol) Limited

Printed and bound in Great Britain by
Biddles Ltd, Guildford and King's Lynn

British Library Cataloguing in Publication Data
A catalogue record for this book is available from the British Library

Library of Congress Cataloging in Publication Data
Rodaway, Paul
Sensuous Geographies : Body, sense, and place / Paul Rodaway.
p.   cm.
Includes bibliographical references.
1. Geographic perception.   2. Senses and sensation.   I. Title.
G71.5.R63   1995
304.2—dc20   93-40331

ISBN 0–415–08829–1

CONTENTS

ILLUSTRATIONS

FIGURES

TABLES

PREFACE

Sensuous Geographies arises out of a long-term fascination which I have had with the role of the senses in the human experience of the environment.

It is difficult to identify the precise genesis of the book, but three interests probably contributed most. First, teaching a new course called 'Individual Geographies' on the geography of children, women, the disabled and the elderly, both at the level of social experience and individual perception, I became increasingly aware of the richness of these 'hidden' geographies in terms of the use of the senses (sight, hearing, touch, smell) and in the role of the emotions, and a general lack of a contemporary general text which could anchor the student's studies and give them a sense of a wider whole. *Sensuous Geographies* does not deal with the social issues of individual geographies, but does explore some of the dimensions of individual perceptions.

Second, I have for a long time found perception studies in geography too imitative of other disciplines, especially psychology, and lacking specific attention to the immediate role of the different senses in generating senses of place and orienting us in space. Nevertheless, I found Gibson's 'ecological theory of perception' most thought-provoking. Whilst formulated in visual terms, it had general applicability to all the senses. He argued that in addition to considering the sense organs and cognitive properties of the brain, we also must recognise the way in which the senses operate as a part of perceptual systems inclusive of the body, its muscles and locomotion, and that all the senses work in close inter-relationship with one another. Furthermore, and most importantly for me at the time, Gibson's theory asserted that the environment itself, the context of perception, played a vital role in structuring stimulation received by the senses – reflected, echoed, disseminated. *This was a geographical theory of perception.*

Third, I have for several years taken a keen interest in the *postmodern challenge*. In particular, I have been drawn to the debates about the body, about representation and on the nature of reality (hyper-reality) in contemporary culture and economy. In writing the present book, I have been keen to take on board some of the ideas of postmodernism concerning the senses and the challenge of living in a world dominated by advanced technologies of mass media consumer cultures and information technology economies. This led me to wonder how have these rapid cultural and technological changes altered the way in which we employ our senses and interpret the information they generate?

Sensuous Geographies is not a comprehensive text, nor as thorough as I would like, but it provides an introduction to a field of interest which only a limited number of geographers have attended to. It bears some relationship to perception geography, humanistic work and recent discussion of the postmodern in geography, but it lays no claim to fit neatly into any of these camps. Instead, it suggests a kind of intimate geography; one which begins with the senses and reaches out to questions of sense and reality.

Paul Rodaway
Edge Hill College

ACKNOWLEDGEMENTS

This book would not have been possible without inspiration given to me by Dr Douglas Pocock whilst I was at Durham University, the support of colleagues at the West Sussex Institute, my family and the Routledge editor, Tristan Palmer. I am also grateful to Ann Chapman of Edge Hill College for preparing the artwork. This book has been long in the making and will continue to be re-made in subsequent attempts to understand the sensuous dimensions of geographical experience.

I would like to acknowledge the kind permission given to reproduce copyright material in various figures. Specifically, Oxford University Press for kind permission to reproduce material as Figure 2.1; Professor F.H. George for the reproduction of the 'range of the senses' as Figure 3.1; The Longman Group, UK for Figure 4.4; Penguin Books Ltd for Figure 5.3; Cambridge University Press for Table 6.1; The British Library for Figure 7.5.3 (reference RR7494); and the London Transport Museum (LTR Registered user No 93/E/596) for Figure 7.5.5.

Part I

SENSE AND GEOGRAPHY

1

SENSUOUS GEOGRAPHY

Words are like nails. You can bang them in like nails and can try to pull them back, as if retrieving them from the wood with the reverse of the hammer, but always they leave holes.

(Anon. 1990)

INTRODUCTION

'We learn to see a thing by learning to describe it' (Williams 1965: 39). *Sensuous Geographies* seeks to describe some of the features of a geography of the senses and to offer some possible explanations of the changing role of the senses in everyday experiences of space and place. This is a surprisingly neglected geography and one rich in possibilities for meaningful investigation (see Ackermann 1990).

Richard Long (Long and Cork 1988), the sculptor, artist and walker, recently argued that: 'I would like to see art as a return to the senses'. Porteous (1986b) has also argued for a return to more 'intimate sensing' in geography as a complement to the current widespread interest in techniques of remote sensing. 'We are far more out of touch with even the nearest approaches of the infinite reaches of inner space than we now are with the reaches of outer space ... what would happen if some of us then started to see, hear, touch, smell and taste things?' (Laing 1967 in Porteous 1990: xvi). The senses are an important part of everyday experience, not just art, providing us with both information about a world around us and, through their structure and the way we use them, the senses mediate that experience. The sensuous – the experience of the senses – is the ground base on which a wider geographical understanding can be constructed.

Richard Long, therefore, builds his art from walking across the landscape, from an intimate sensual experience with space and the materials of his environment and forms his 'sculptures' in pattern with the landscape, its structure and material substance, both by leaving arrangements of stones and other materials in lines or circles in the landscape, and photographs of these and material versions compiled in exhibition galleries (Romey 1987; Long

3

and Cork 1988). Geographers, perhaps, need to return in some way to a kind of sensual study, both intimate in its focus on the information of the senses – touch, smell, taste, hearing, sight – and also wider ranging, inclusive not just of the visual dimension of experience, but also the other senses. Sullivan and Gill have observed: 'sight paints a picture of life, but sound, touch, taste and smell are actually life itself' (1975: 181). A sensuous geography may therefore lay some claim to reasserting a return of geographical study to the fullness of a living world or everyday life as a multisensual and multidimensional situatedness in space and in relationship to places.

Sensuous Geographies is an exploratory study. Its primary aim is to excite interest in the immediate sensuous experience of the world and to investigate the role of the senses – touch, smell, hearing and sight – in geographical experience. The potential scope of such a text is vast and thus, inevitably, the survey is selective. These four senses are selected as the most immediately relevant to geographical experience and each is considered in turn but this is not to deny the multisensual nature of everyday experience and the value of considering other sensual dimensions of geographical experience.

The sense geographies identified are convenient analytical categories designed to focus attention on the specific qualities each sense gives to geographical understanding and do not suggest independent or alternative worlds. Nevertheless, the senses are not merely passive receptors of particular kinds of environmental stimuli but are actively involved in the structuring of that information and are significant in the overall sense of a world achieved by the sentient. In this way, sense and reality are related.

Interest in sensuous geographies is not new. The aesthetic geography of Vaughan Cornish (1928, 1935), the perception geography of the late 1960s and early 1970s (Lowenthal 1961; Kirk 1963; Gold 1980) and the more recent work of humanistic geographers (Pocock 1983, 1989, 1993; Porteous 1985, 1986b, 1990; Porteous and Mastin 1985; Tuan 1974, 1993) have each brought attention to human perception in geographical understanding. The present book, however, focuses exclusively on the senses and geographical experience.

A number of methods of investigation are possible: a scientific or psychophysical analysis of the spatial dimension of sense perception (e.g. Blauert 1983), a sociological analysis of the socio-historical definition of styles of sensuous experience (e.g. Rose 1986), a cultural analysis of place design and sensuous experience (e.g. Norberg-Schulz 1980). Each approach provides important insights but they all tend to separate out the physical, social, cultural and aesthetic dimensions of human experience. The present study seeks to offer a more integrated view of the role of the senses in geographical understanding: *the senses both as a relationship to a world and the senses as in themselves a kind of structuring of space and defining of place.* Therefore, the current study is more eclectic in method, drawing on each of these perspectives in conjunction with phenomenological reflection and the

4

introduction of ideas from recent postmodern writing.

Of course, the very act of focusing on the senses is full of presuppositions and constitutes an abstraction. It presumes that distinctive senses can be identified and that their role in geographical experience can be discussed meaningfully individually and separate from the emotional dimensions of experience. Sensual and emotional geographies are closely connected, as is well demonstrated in the writings of Yi-Fu Tuan. However, here we concentrate on sensuous geographies and leave emotional geographies for another project.

Everyday experience is multisensual, though one or more senses may be dominant in a given situation. These abstractions are an analytical device to enable us to highlight often taken-for-granted and hidden dimensions of geographical experience. Here, geography is understood as earth (*geo-*) drawing (*-graphe*), that is, a description of the earth and human experience of it, considering issues of orientation, spatial relationship and the character of places. *'Sensuous geography' therefore refers to a study of the geographical understanding which arises out of the stimulation of, or apprehension by, the senses.* This is both an individual and a social geography, a physical and a cultural geography.

SENSE AND SENSES

This is not a perception geography, nor an experiential geography, but a geography of the senses. The term 'sense' has an important duality or ambiguity.

1 Sense, as in *'making sense'*, refers to order and understanding. This is sense as meaning.
2 Sense, or *'the senses'*, can also refer to the specific sense modes – touch, smell, taste, sight, hearing and the sense of balance. This is sense as sensation or feeling.

These two aspects are closely related and often implied by each other. The sense(s) is (are) both a reaching out to the world as a source of information and an understanding of that world so gathered. This sensuous experience and understanding is grounded in previous experience and expectation, each dependent on sensual and sensory capacities and educational training and cultural conditioning.

The ambiguity of the term 'sense' – referring to specific sense organs (sensation) and broader mental constructs (meaning) – is also a relationship between the immediate experience and metaphorical extrapolation. This metaphorical dimension has been explored by a number of geographers (Tuan 1979a; Pocock 1981a; Porteous 1990). The reason for metaphorical uses of the senses lies, in part, in the multisensual nature of everyday geographical experience and the complex and ambiguous relationship between the individual senses. Synaesthesia – secondary sensation caused by stimulation of another part of the body, such as accompanying sensations of colour with given sounds – is a related phenomenon, but here the more basic inter-relationship and cooperation of the senses in everyday experience is meant. Metaphorical use is also reinforced by language traditions, especially in Western cultures, in which visual metaphors are often preferred (Pocock 1981a) and vocabularies of visual and, to a slightly lesser degree, auditory and tactile experience are rich, whilst those for smell and taste are more limited and dependent on associations and metaphors of the other senses. Sense is sensation and meaning and, therefore, the term 'sense' – literal and meta-phorical – leads to deeper questions about sense and reality. Different cultures in different times and places have employed or defined the senses differently and their associated concepts of reality. Therefore, a sensuous geography cannot just describe the experience of the senses and their role in the constitution of geographical experience, it must also consider more fundamental questions about the nature of person–environment relation-ships and what constitutes a geographical reality for a given society (or culture) at a given moment in time and space.

GEOGRAPHY, HUMANISTIC OR POSTMODERN?

Sensuous Geographies combines ideas from both humanistic and postmodern thought. At the end of the 1960s, human geographers became somewhat disillusioned with the so-called 'quantitative revolution' and the methods of positivist science. Geographers studying perception and behaviour were especially concerned about the important qualitative dimensions of human experience which such aggregate methods missed. There was also a revival of interest in human geography as a whole in the issue of relevance and the need to solve 'real problems' for society and the environment. In the 1970s, therefore, a number of competing approaches or new philosophies of geography emerged (Johnston 1983a, 1983b; Buttimer 1993). In human geography, the main alternatives were a form of Marxist social geography grounded in ideas from sociology and political economy and a humanistic individual geography which related more to the humanities and arts. In a 'battle' for dominance in the discipline, Marxist geography seemed to triumph over the humanistic approach and the latter became the preserve of a few dedicated researchers (notably Yi-Fu Tuan).

More recently, and especially since the mid-1980s, geographers have re-evaluated their reading of sociology, adopting a broader range of social theories and journeying into the debates about postmodernism (Dear 1988; Soja 1989; Harvey 1989). At the same time, there has also been a revival of interest in humanistic issues and some cross-fertilisation between humanistic and postmodern thinking in geography (Tuan 1993; Buttimer 1993). The present book is situated within this context of a reading of postmodernism and a re-evaluation of humanistic perspectives within the framework of a study of the senses and geographical experience.

Within the complex debates about postmodernism and the contemporary social and geographical experience, one can identify three key issues which are each of immediate relevance to the definition and scope of a sensuous geography.

1   *The redefinition of the 'real' and the position of the 'sign'.* This is the question of reality versus representation, of sign and referent, and is typified by Baudrillard's concept of hyper-reality (1983a). Here, the senses as an access to a 'real world' are not naively defined, that is, not merely as physically absolute characteristics, but are situated within and defined by the socio-economic, cultural and technological context of their employment.

2   *The reassertion of 'space' in social thought,* or specifically the role of spatial (and space–time) frameworks (or logics) in analysis. Social and geographical understanding is always and already situated within specific places and times, and an understanding of both the specific and the general must take account of the complex inter-relationships or networks of past, present and potential operating through a given space. Bakhtin's concept of the chronotope (1986a, 1986b; Folch-Serra 1990) and Foucault's dispositif and detailed studies of the transition to modernity represent this spatialised thinking (Foucault 1970, 1972, 1973, 1975, 1979, 1967/1986; Shields 1991; Philo 1992). Each of these contexts implies a changing sensuous geography.

3   *The rediscovery of the sensuous (and the body)* as a potent part of social, political, historical and geographical experience. This introduces the discussion about rationality and the economy of desire (Lyotard 1971; Deleuze and Guattari 1984) and reasserts the body as a focus of socio-political struggle and cultural change. The senses, situated on the body and operating through the body, and the body itself as a sensuous dimension, gain new significance in social and geographical understanding.

Lash (1988) argues that (in the cultural realm) modernism is discursive and textual, whilst postmodernism is figural and sensory. The first is about meaning, the second is about experience. The humanistic perspective and, specifically, the phenomenological approach to perception (Merleau-Ponty 1962) asserts a unity of experience and meaning, that is sense(s) is (are) both sensation and meaning. The re-evaluation of humanistic approaches and specifically phenomenology (Seamon 1979; Seamon and Mugerauer 1985) indicate a number of other key issues for a sensuous geography.

1   *Phenomenological reduction or 'back to the things themselves'* (Husserl 1983). Heidegger describes the phenomenological attitude: 'to let that

which shows itself be seen from itself in the very way in which it shows itself from itself' (1983: 58). Whilst all the senses form part of a multisensual whole, each can be observed individually to appreciate its distinctive 'voice' or role in geographical experience and studied to chart its distinctive 'history' and changing place in geographical under-standing.

Johnson (1983) identifies three stages of reduction in Husserl's phenomenological strategy: phenomenological reduction – suspension of all beliefs characteristic of the 'natural attitude', that is, everyday common sense and the scientific method – this is a 'bracketing out' of preconceptions; eidetic reduction – when the particular encounter with the phenomena is taken as universal; and, thirdly, a kind of psychological reduction by which we discern the constitution of phenomena in our cognition. (Postmodern cultural analysis or readings perhaps offer a more open and accessible form of reduction. Derrida's (1982) decon-struction could also be interpreted as a kind of reduction in this sense.)

2   *Intentionality and anthropocentrism* – phenomenology begins and ends with the human subject, seeking to understand the nature of 'being in the world' (Heidegger 1983) and the recurring theme of phenomenology, according to Husserl (1983), is intentionality and specifically human intentionality. Our experience – including sensuous experience – is always and already a consciousness of something, and we have a relationship to that thing, it participates with us in constituting a world. The intentionality of experience is therefore a sense of ownness and belonging, of relationship and participation, of situatedness. Phenomen-ology implicitly gives geography and its fundamental question 'where?' a key place in the understanding of human experience.

3   *Wholeness and participation* – phenomenology is grounded in the realisation that we are already within and part of the world we study. It is not possible to sustain an objective and detached view of the world. Geographical understanding always begins from or is relative to a given location in space, the space which is being studied. Further, the act of studying inevitably effects or changes the thing studied and the individ-ual studying it. There is an interdependence of phenomena studied and methods (Seamon 1983). Therefore, through reflection on sensuous experience and the study of sensuous geographies, we each become both more aware of our own sensuous geographies but also those experiences are changed.

The phenomenological attitude has been variously equated with 'meditative thinking' (Heidegger 1966), wonder (Fink 1933; Relph 1985), opening (Giorgi 1970), surrender (Wolff 1963), spiritual discipline (Zimmerman 1985) or love (Laing 1964). It is a kind of heightened experience, more aesthetic than immediately practical.

The present text seeks to combine elements of humanistic and postmodern thinking, taking a journey from phenomenologically inspired reflections on the nature of each sense dimension and its associated geography and going on to explore the cultural definition of that sense and the contemporary experience of sensuous reality (see also Shields 1991).

A GEOGRAPHY OF THE SENSES

The present text falls into three parts:

1   *Sense and geography* – considers the roots of sensuous geography in perception/behaviour and humanistic geographies, and the general character of the senses.
2   *Sense, space and place* – considers four senses (touch, smell, hearing and sight) in turn, identifying their distinctive contributions to geographical experience at individual and social levels, in different historical, cultural and technological contexts. The examples are illustrative and not exhaustive and by focusing on each sense in turn it is not intended to deny that everyday geographical experience is both multisensual and emotional.
3   *Sense and reality* – returns to theoretical issues and makes a number of speculations about the relationship between styles of sensuous experience and concepts of geographical reality.

Much of the humanistic analysis of sensuous geographies has been covered in recent books by Porteous (1990) and Tuan (1993) who each identify the importance of both literal and metaphorical sensuous geographies in aesthetic and everyday experience in Western and non-Western cultures. The originality of the present study lies in combining reflections on the geographical nature of each of the four senses – touch, smell, hearing and sight – and their role in different cultural contexts, and in developing the literal/metaphorical distinction by employing ideas from the writings of Jean Baudrillard on human experience, reality and representation (1983a, etc.). In particular, his notion of the orders of simulacra and the concept of hyper-reality are adopted to explain the socio-historical development of styles of sensuous experience and the consequent changes in concepts of reality through the introduction of new social practices and the employment of new technologies. The text progresses from basic issues of definition, through illustrations of the variety of sensuous geographies, to more general issues of concepts of geographical reality. The book, therefore, aims to challenge the reader to reconsider the role of the sensuous, not merely as a physical basis of geographical understanding but also as an integral part of the cultural definition of geographical knowledge.

2

PERCEPTION THEORY AND THE SENSES

PERCEPTION AND SENSE

Perception is a word which defies precise definition. *Percipere* (Latin) denotes: 'to take hold of, to feel, comprehend'. From this root, Tuan (1974) suggests that 'perception is an activity, a reaching out to the world'. However, perception has many meanings, technical and everyday, each of which play on any attempt to hold to the simple statement that 'perception is ...'. This has strongly influenced how researchers have posed the question of perception in research design and theory development.

Two salient connotations are found in everyday usage:

1 perception as the *reception of information* through the sense organs associated with sight, hearing, touch, taste, smell; and
2 perception as *mental insight*, or a sense made of a range of sensory information, with memories and expectations.

The first is perception as sensation or feeling; the second is perception as cognition or insight. The two connotations are not mutually exclusive, but each implies the other. This duality mirrors that we have already identified in the term 'sense' – sensation and meaning. To this extent, sense and perception might be exchangeable terms, although the latter is more precisely an activity or process. Therefore, Allport (1955: 14), for instance, adopts the frequent general definition of perception – the 'catch-all':

> perception has something to do with our awareness of the object conditions about us, it is dependent to a large extent upon the impressions these objects make upon our senses ... But perception also involves, to some degree, an understanding awareness, a meaning or a recognition of these objects.

In scientific use, the term 'perception' is generally given a more limited and precise meaning, one which reflects a specific theory of perception. Such definitions are often at the cost of emphasising one dimension over the other;

10

physical sensations or cognitive processes. However, one common characteristic of these more specific definitions is that they generally emphasise
perception as a process, an activity involving the organism and its environment.

An effective geographical understanding of perception needs to recognise
both dimensions of the term:

1   *perception as sensation*, and therefore a relationship between person and
    world, both kinetic and biochemical (here perception is grounded in the
    environmental stimuli collected – and mediated – by the senses); and
2   *perception as cognition*, and therefore as a mental process (here percep-
    tion involves remembering, recognition, association, and other thinking
    processes – which are culturally mediated).

The consequences of understanding perception as a complex of both these
dimensions is really to recognise that we do not perceive naively. Our
perception is influenced both by the efficiency of our sense organs (which
differ from individual to individual) and by our mental preconceptions
(individual training, cultural conditioning). In this formulation perception is
grounded in sensations which are a series of environmental stimuli and
involves cumulative, analytical and synthetic, processes of the brain, each
working together to give us a sense of a world, or geographical understanding. *Perception is therefore a relationship to the world and a decision-
making process with respect to that world.*

However, this simplified formulation omits important dimensions of the
process of perception. It focuses primarily upon the mechanism of perception and ignores the wider context of perception. A fuller definition of
perception needs to include four further dimensions.

First, perception is not the reception of single stimuli from one source
direct to the sense organ concerned, but rather involves a myriad of different
stimuli from various sources reaching the different sense organs. This
includes sensuous abundance, ambiguity and redundancy. Furthermore,
whilst it is common to refer to visual perception and auditory perception, for
instance, actual everyday perception is characteristically multisensual,
involving more than one sense organ in generating an experience of the world
(Tuan 1993). Therefore, it is important to consider the way in which the
senses interact, as well as the operation of specific sense modes.

Second, perception is not a direct, or isolated, process but the interaction
of stimuli, sense organs and the brain, takes place within an environment, or
geographical context. The 'space' across which a stimulus crosses from
source to sense organ is not empty, it is made up of different surfaces, textures
and objects which can interrupt the passage and/or alter in some way the
nature of the stimulus message. This is sometimes known as an ecological
concept of perception (see pp. 19–22).

Third, perception is a learned behaviour. Therefore, a definition of perception must take account of habituation to a given stimulus (which may reduce its effect) and increased sensitivity to another stimulus (usefully through training). Furthermore, any definition of perception will be culturally specific, the style of perception adopted is the result of a process of socialisation.

Fourth, perception is corporeal; it is mediated by our bodies and the technological extensions employed by the body (such as walking sticks, spectacles and hearing aids, and even clothes). The body is more than the site of the sense organs and the brain, but forms a fundamental part of the perception process. Its size and orientation, its locomotion and its own sensuous capacities (balance, for instance) are important issues for perception.

Since the seventeenth century, there has been a persistent tendency to distinguish between sensations and perception (Gibson 1968). They have been placed in a kind of hierarchical dichotomy: sensation as inferior, primary raw data, and perception as superior, interpretation and knowledge. Gibson argues that this taken-for-granted distinction, or presupposition, has prevented the development of an effective, comprehensive theory of perception. Many theories of perception can explain simple laboratory style situations but not the more complex day-to-day geographical experience of a world. The nativism–empiricism debate – that is between innate or inherited knowledge and knowledge learnt through experience – and the more advanced theories of behaviourists, cognitivists and Gestalt psychology, all fail to break out of this sensation/perception framework. This presupposition repeats the age old mind–body conflict, one in which thinking matter and material substance are juxtaposed as determining and determined, or active and passive. Yet, perception – as a combination of sensation and cognition – is inclusive of both passive encounter with environmental stimuli and active exploration of that environment, as the body moves through space and time interacting with a world.

Perception is not just sensation plus cognition since human beings each have a personal history (biography) and are socially and culturally situated. Individuals in different societies differ in what they perceive and value in their perception. This is most starkly observed when comparing modern West European and North American perceptions with those of other traditions, notably Arab, Japanese and various aboriginal groups (Hall 1969; Carpenter 1973; Lopez 1986). Even within a given society, there are marked difference in what individuals of different ages and sexes, education and training, and socio-economic background perceive and value in their perceptions. Therefore, a definition of perception must not only be grounded in psychophysical theories of perception but also in socio-historical theories of the emergence of specific styles or priorities in human perception. The importance of the visual, for instance, in contemporary experience (from everyday language to technological advances) is not merely the accident of biological evolution but more likely the consequence of a specific socio-historical development (see Cosgrove 1984).

Therefore, perception involves the sense organs (including the body) and the mind, but is also situated in and mediated by a geographical and cultural environment.

GEOGRAPHY AND PERCEPTION

Geography is concerned with both a physical and human world. Literally, geography is earth ('*geo-*') drawing ('*-graphe*'). It is a representation of an experienced world, both perceived directly through the senses and mediated by the mind. Perception as a process by which data is collected and ordered is therefore fundamental to geographical enquiry. Geographical perception is simply the perception of a world around us, of spatial relationships and the identification of distinctive places – to recognise our situation in a world and to have a sense of a world. Sensuous geography – as a focus on the role of the senses – might be located in the sub-field of perception studies.

In general, geographers have been somewhat uncritical in their use of the term 'perception' and have drawn heavily upon a limited range of psychological theories whilst not generally taking account of the critical traditions underlying these theories. A move from a psychological and biological interest in perception processes to a geographical one is far more fundamental than was originally realised. Geographical interest is in perception in complex everyday environments and decision-making in widely differing socio-economic and cultural contexts. These wider issues of perception call for a distinctive geographical formulation of perception models.

There have been numerous reviews and collections of 'perception geography' (e.g. Wood 1970; Gold 1980; Saarinen *et al.* 1984; Aitken 1991). Some reference has been made to ethnocentric and eurocentric views of the world (e.g. Tuan 1974), but in general this dimension has been given little attention. The behavioural or perceived environment (Kirk 1963) has been, with hindsight, perhaps the most influential concept to develop out of the interest of geographers in perception. This suggests that individuals and groups do not make decisions based on full, accurate and objective knowledge of their environment, but a more partial, less accurate and subjective view of their environment derived from the mediation of previous experience, education and social conditioning, and expectations and hopes. Behavioural decisions are made on the basis on this environment as perceived (Brookfield 1969) rather than the actual environment, and as a consequence of the mismatch of perceived and actual worlds mistaken behaviours occur. This interpretation of perception, however, tended to equate perception with subjective experience and reduced the perception question to a behavioural one of intervening factors, such as community decision-making processes and the dissemination of environmental information. The more direct interest in perception as a

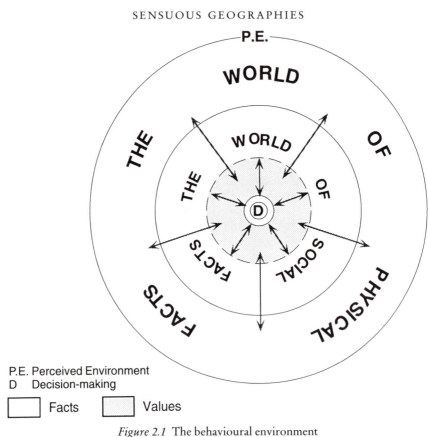

P.E. Perceived Environment
D Decision-making

☐ Facts   ▨ Values

Figure 2.1 The behavioural environment
Source: Kirk (1963) in Jones (1975: 99) by permission of the Oxford University Press

mode of encounter with the world, as the process of acquiring knowledge about a world, was given much less direct attention by geographers.

In the late 1960s and early 1970s the sub-field 'perception geography' (sometimes called 'behavioural geography') gained much prominence. This sub-field drew heavily upon positivist psychology, chiefly various behavioural and cognitive theories of perception (see Seamon 1979). In the late 1960s and early 1970s the humanistic and phenomenological traditions of psychology were largely ignored in favour of psychology (and geography) as social (and even physical) science. This was not, however, the first occurrence of perception studies in geography. Pioneer work can be traced back to the early years of the century and in Britain the work of Cornish (1935) on scenery and the sense of sight is notable. This early work was more directly inspired by aesthetics and the experience of art, and was grounded in a concept of perception as the process of direct experience of the world through the senses and interpreted by the mind.

These aesthetic and psychological approaches to perception differ in a number of ways. The aesthetic approach takes a more direct and naive view of perception as sensuous experience and mental reflection equating this to a precise and objective knowledge (see Cornish 1935). The psychological approach takes a more indirect interest in effects, or behaviours and/or mental constructs, arising from encounter with environmental stimuli, accounting for intervening factors (such as education and training) and emphasises the subjectivity of human perception – as in the behavioural environment concept. Yet each tradition focused upon perception as an individual encounter with the environment, one which was largely passive and predominantly a visual experience. Even later cognitive models, such as the mental map (Gould and White 1974), placed emphasis on visual cues and map-like images. The wider sensual and emotional composition of human perception was given much less attention by perception geographers. More recently, since the mid-1970s humanistic geographers have begun to revive interest in the aesthetic – sensual and emotional – dimensions of geographical experience and a more cultural interpretation of perception has emerged (Tuan 1974, 1979b, 1993; Seamon and Mugerauer 1985). This alternative interpretation of perception – or, more correctly, human–environment encounter (e.g. Seamon 1979) – draws on a phenomenological tradition (such as Merleau-Ponty 1962) and considers both the direct sensuous experience of environments and descriptions of places in literature and art (Pocock 1981b).

Geographical understanding of perception it seems has two epistemological dispositions:

1   *idealist* – emphasises the perspective of the individual – seeing the world through the eyes of another person; and
2   *positivist* – models of perception as a mechanism of stimulus–response (behavioural) or as information-processing mechanisms and mental constructs (cognitive).

Since the late 1970s, perception geography has declined in importance and been marginalised. This change is in part the result of the failure of perception geography itself – and in particular its adopted models of perception – and the result of wider changes in the discipline of geography as a whole – especially the rapid growth of sociologically inspired geographical theory and the development of phenomenologically inspired humanistic geography.

The concept of perception dominant at the height of 'perception geography' (approximately 1965–75) was strangely contradictory and constrained by a number of rarely discussed presuppositions. For instance, the idealist/positivist dispositions were potentially contradictory. Idealist perspectives emphasise the individual and a radical subjectivism. This perspective tends to assume a passive or contemplative observer more involved with art appreciation than participation in an everyday world. This was particularly evident in landscape perception and preference studies (Mitchell 1979; Punter 1982).

Positivist methodologies attempt objective quantitative measurement of stimuli and identified behavioural responses, and to model the general character of perception and behaviour (and, in the case of cognitive approaches, mental processes). Positivists seek to sample populations and ultimately to recognise general laws for the whole population. It is assumed that individuals are rational (usually male) observers and behave in predictable and repeatable ways.

Recent humanistic geography has revived interest in individual experience, but drawing more upon a phenomenological perspective, and taking greater account of cultural differences in perception. This arose in part from a disillusionment with more positivist methodologies, especially when dealing with complex qualitiative values, and in part from the rediscovery and reassessment of work in the humanities and cultural studies, as well as humanistic, phenomenological and interpretive methodologies. As Tuan writes: 'the power of the human senses to organise the world takes diverse forms, shaped by their larger cultures in which they operate' (1993: 122). An alternative view of perception in geography is emerging to meet this challenge, one which draws upon psychophysical, aesthetic and cultural perspectives.

PSYCHOLOGICAL MODELS

'Classical' perception geography (*circa* 1965–75) subscribed to one of two largely positivist theoretical traditions in psychology – the behavioural and the cognitive. These are not single theories, nor are they mutually independent, since during the course of research numerous model variants of each were developed and, increasingly, attempts were made to combine elements of both in a single model. Space does not permit us to explore the complex debates which raged in psychology over these two theoretical perspectives, but it is important to identify the key positions each represents since they were so important in shaping geographical concepts of human perception. This and the following section offers a brief overview of three alternative pyschological models of perception – behavioural, cognitive and ecological. This is highly selective and omits other important perception models such as transactional models (Ittleson 1974), phenomenological perspectives (Merleau-Ponty 1962; Ihde 1976) and insights from Gestalt theory. However, this section reflects the history of perception models in geography and the following section explores a more appropriate alternative, Gibson's ecological model.

The label 'behavioural models' refers to a range of variations on the stimulus–response model made famous by Pavlov's dogs (e.g. Murch 1973). They range from highly mechanistic models of a specific stimulus leading to a specific behavioural response, to complex models of multiple and alternative stimuli and associated complex behavioural scenarios involving

16

concepts such as thresholds (levels of stimulus necessary before expected behaviour occurs), habituation (reduced response after repeated exposure to a stimulus) and chance. Some models have emphasised the role of instinct – pre-programmed responses to given stimuli. Other models have emphasised the importance of learning in stimulus response.

Research in this tradition – both on animal and human perception – has often been highly focused and conducted in controlled laboratory conditions. Specific hypotheses are formulated concerning particular stimuli, intensities and repetition of stimuli, and expected responses or behaviours. The behavioural models were largely concerned with inputs and outputs and not the physical and mental processes in between. This has led to the charge that such models of perception were mechanistic and 'black-box'. Even when behavioural responses to a given type and intensity of stimulation is successfully predicted, it remains a puzzle why or how the result occurs, and it remains largely a laboratory abstraction and less predictable in the complexities of the outside world. In the lived world not only are stimuli multiple but also many contextual issues influence how stimuli are responded to. Most behavioural models could not take full account of this level of complexity.

Hindsight makes behavioural models appear quite naive but we must remember that they contributed much to our understanding of stimuli and stimulus–response associations, the complex of instinctual and learnt behaviour, and introduced a number of valuable concepts such as habituation and thresholds. Nevertheless, behavioural models were ultimately constrained by a number of presuppositions as well as the omissions already noted. In particular, a remarkably simple model of cause and effect was assumed. Even in later multiple stimuli/complex response models, causality was essentially uni-directional from environment to behaviour, with the former reduced to recognised stimuli and the latter to specific measurable responses. The perceiver was largely passive and did not appear to have the freedom to make choices, rational or irrational. The idea of search behaviour, that is, active perception and interaction with the environment – as developed in transactional models (Ittleson 1974) – was slow to emerge in behavioural writing. Despite a gradual shift from reflex and instinctual models to learning and habituation models, causality remained naively formulated and above all the perceiver was assumed to be a rational mechanism, an optimising organism. Behavioural models gave little room for irrational behaviour, for freedom to choose or make mistakes in perception. Perception was a rational process and essentially rooted in biological needs – either immediate needs or less directly a legacy of evolutionary conditioning (Appleton 1975). Wider human perceptual behaviour, such as artistic contemplation, was ignored or fitted awkwardly in this tradition (Wohlwill 1976).

Cognitive approaches, in part, grew out of a critique of the 'black-box' mechanism of behavioural models and reflect a shift of interest towards the role of learning in perception and interest in decision making processes. Cognitive models do not dismiss the role of physical stimuli but put greater emphasis on mental processes – that is, the interpretation of sensuous information, analysis and synthesis by the brain. In this sense, cognitive models take a more active view of perception since cognition is defined as thinking, figuring out or deciding. Mental processes are the core interest but, inevitably, precise observation and description of specific mental processes proved difficult. Different analogies, or competing concepts of the brain, distinguish the various cognitive models (e.g. Downs and Stea 1973; Moore and Golledge 1976; Roth and Frisby 1986).

Analogy is both the basic strength and weakness of cognitive models and signals an important shift in scientific approach. Cognitive models essentially propose particular concepts of mental processes and, by manipulating stimuli and monitoring responses (including in some cases brain activity), hope to evaluate the accuracy or coherence of such concepts. The most widely explored models were based on an analogy to the computer or 'information technology' (see Roth and Frisby 1986). The brain is hardware and particular patterns of thinking or mental constructs represent software programmes (either instinctual or established by learning). The environmental stimuli remain the raw data of sensations from the sense organs but are now called information or inputs. This is a little confusing when compared to Gibson's ecological model of perception (see next section). In that model, raw sensations are not perceived (or understood) but rather stimuli which have been structured by the environment through which they pass and thus acquire information. In cognitive models the cause–effect model remains largely unchanged but now the '–' between is given closer attention. Mental faculties such as memory and recall, recognition and expectation, and choice between alternative information sources, become possible research questions for cognitive theorists.

Seamon (1979) identifies two cognitive positions. First, those grounded in the behavioural tradition which continue to accept the stimulus–response model (Tolman 1973). Here perception is still largely a passive process. Second, others take a more 'interactive-constructivist' position. This has been linked to the ideas of Piaget (Hart and Moore 1976). This approach argues that the person actively mediates his/her relationship to the environment (Piaget and Inhelder 1956).

It is out of the cognitive tradition that mental maps (Gould and White 1974) and personal constructs entered geographical thinking. The mental map or visual image analogy of information stored in the brain has been particularly influential in geography. This analogy, however, tended to give a static rather than dynamic view of the mental dimension of perception. The cognitive map is essentially a mental image of spatial relationships held by the brain which, it is proposed, guides way-finding behaviour. Much research has been done on mental maps, especially with children (Matthews 1980,

1984; Blades and Spencer 1986, 1988; Spencer and Darvizeh 1983). However, it has not been possible to prove whether such 'maps' exist (or other mental constructs) and how they actually relate to the representations drawn or described. As Tuan put it, 'it cannot be assumed that people walk around with pictures in their heads, or that people's spatial behaviour is guided by picture-like images and mental maps that are like real maps' (1975: 213). Nevertheless, cognitive models have proved more productive for geographers than behavioural models due to this greater attention to spatial concepts and problems such as navigation in the environment.

Like behavioural models, cognitive models have a number of presuppositions. Perception is still assumed to be a rational process and, despite a greater emphasis on learning, cultural and social dimensions remain secondary. Much of the research had a visualist emphasis, most notably in the notion of 'mental image'. The senses remain the sources of the raw data of sensation and the brain is seen as the sole structuring or interpretive agent. The context of perception, the complex matrix of stimuli in a given situation and the wider issues of total body sensitivity are not explored in cognitive models.

Cognitive models still have much impact on perception research, especially attempts to develop artificial intelligence systems or computerised robots (Roth and Frisby 1986). Object recognition and questions of environmental navigation remain problems of information processing in the brain and specifically mental organisation.

AN ECOLOGICAL MODEL

The sense of a world given by perception has been attributed to the performance of the senses themselves (physiological theories), explained by activities of the brain (psychological theories – implicitly in behavioural, explicitly in cognitive approaches) – and considered a result of styles of human–environment encounter (phenomenological and transactional theories). However, Gibson (1968, 1974; Heft 1988) argues that the structure and texture of the environment itself is a necessary determinant of what is perceived. This 'ecological' theory, as he calls it, is of special relevance to geography, since it not only gives importance to the environment itself in perception but also considers perception by a mobile observer. Gibson's initial formulations of the theory were for visual perception and specifically spatial perception (of arrangement, depth and movement) by aircraft pilots (Gibson 1974), but the basic principles of the theory can be applied to all sensuous encounter with the environment.

Gibson abandoned the traditional sensation/perception distinction and transcended the behaviour/cognition debate and presents a theory which attends specifically to spatial perception and the active observer moving in a dynamic world. The ecological theory of perception has two key features.

First, the senses are considered as *perceptual systems* – this emphasises the

19

inter-relationship between the different senses (including the body itself) in perception and the integration of sensory, bodily and mental processes (see p. 29).

Second, the concept of *ecological optics* (and ecological formulations of other sensuous information) emphasises the role of the environment itself in structuring optical (auditory, tactile, etc.) stimulation. Potential sources of stimulation pass through the environment and are encoded with the structure of that environment as they are modified in their passage. It is this structured stimulus which the sense organs 'read'. Therefore, the environment becomes a source of *information*, not merely raw data.

In contrast to previous psychological models, Gibson's approach explores space perception in total environments rather than in point-stimulus situations developed in laboratory conditions. Here, perception study does not begin with a stream of raw data sensations, nor with a priori mental constructs, but with an analysis of environment itself as a complex of surfaces, edges, textures and, importantly, movements.

In identifying the senses as perceptual systems, Gibson (1968) still identifies five broad and inter-related sense systems: basic orientation (associated with the experience of gravity), the auditory system, the haptic (somatic) system (associated with touch), the taste–smell system, and the visual system. These systems are functional and cooperative in an interaction with the environment. Each system is inclusive of both specific sense organs, the associated muscles of the body and mental processes. Stimulation comes not only from the environment but also from the organism itself (Gibson 1968: 31). The internal organs and the movement of extremities, as well as the specifically identified sense organs, and locomotion of the whole body through space (including exploratory activity) provides stimulation. Perception is an experience of the whole body and an activity in a dynamic world. Whilst distinctive perceptual systems are identified analytically, the understanding is ultimately multisensual and emphasises inter-relationships between the sense organs, the body and the brain. Perception is situated corporeally and environmentally, behaviour and perception are implicative of one another.

The environment is not a random set of stimuli but rather sensory information (light, odours, sounds, etc.) is structured (encoded, given meaning) by the environment through which it passes and is received as information about objects and spaces. This is the principle of ecological perception. Gibson (1968: ch. 9) explains this most fully in the context of visual perception of space. He distinguishes between radiant light and ambient light. Radiant light is emitted from an energy source (such as the sun, or a light bulb) and contains the minimum of discernible environmental information. Looking at such light we are blinded by an uninformative whiteness. Instead, we need to pay attention to what is illuminated by that light. Visual perception therefore is derived from ambient light and the eye is a specialised organ for 'reading', or decoding, this. Ambient light is made up of reflected light which is full of information about the structure and properties of the surfaces it is reflected from.

In an illuminated environment, ambient light is structured in a combination of three principal ways:

1   *differential facing or inclination* of different surfaces with respect to the sources of illumination;
2   *reflectance of surfaces* (there are two sorts of reflectance; where reflected light is unselective of wavelength it is a difference in whiteness; where it is selective of wavelength it is a difference in colour);
3   the *different illuminatedness or shadowing* of different parts of a surface.

These allow us to perceive from the structured ambient light, or optic array, the arrangement of surfaces and edges in the environment, the shape and arrangement of objects, and the texture and colour of surfaces. A key feature of the structure of optic arrays is the edge information, the transition from one bundle of reflective rays to another. These edges, or breaks, indicate changes in the slant of surfaces, edges of objects, changes in surface texture and so forth. Different densities of optic arrays indicate depth (or distance) and further spatial differentiation. The specification of a hole in the world, an aperture, a window, another room beyond or space between objects, is that it 'opens up' on an optically more dense array. The movement of the eye and the observer and/or movement in the environment enriches the optical information received by the eye. Exploratory activity thus enriches perception and actual environments are always richer and generally less ambiguous than representations such as a photograph. Equally, stereoscopic vision is richer than monocular vision but it is not necessary to have two eyes in order to perceive depth. The information about surfaces, spatial arrangements and depth is contained already in light received by the eye.

It is possible to explain other sensuous experience in ecological terms. For instance, auditory perception derives from information encoded in sounds by its passage through an environment. Inclination of surfaces and substances towards sound sources, the resonant (or transmission ability) versus absorbent (or reducing) properties of substances (surfaces, objects), and the differing acoustic properties of spaces (volumes), each give the vibrations (or sounds) reaching the ear a specific character and a general kind of 'map' of the environment. This ecological dimension might be called acoustic sound. The emitted sound, in itself, will also probably have certain information encoded in it (and in this sense differs from radiant light in that we can discern it in many cases). The emitted sound, or voice, is modified by the environment it passes through, being enhanced or degraded. The original emitted sound itself can in a certain way also be explained ecologically, though here it is not the environment in general which structures the sound but the vocal chords of the organism or the structure of the objects involved

21

in the generation of the sound. Likewise, touch and smell experience can be explained ecologically.

THE CULTURAL DIMENSION

Perception is a social, or shared experience, as well as an individual one. Whilst individuals differ in the precise details of their perceptions, there is nevertheless an identity or similarity of sensuous worlds shared. Of course, on occasion individuals differ markedly in their perceptions, and this can lead to important misunderstandings and confusions. Those with a sensory disability would appear to have somewhat different geographical experience to the able-bodied (e.g. Hill 1985), some have suggested there are also sex and age differences in perception, people from different professions prioritise different details (and even perceive different details) in their perception, and it is widely recognised that cultures differ in their employment and interpretation of sensuous experience. Underlying all these examples of differences in perception is the idea that, though there may be a certain universal and instinctive level in human perception, by far the most important component is the learnt behaviour. Perceptual sensitivity is learnt and forms part of our socialisation into a cultural group. This is the cultural dimension of perception. Each sense is not only physically grounded but also its use is culturally defined.

Perception may be quite taken for granted in everyday life but it is not just a physical reflex. Perception, or our relative use of the different senses and depth of perception, is a learnt behaviour, a skill. The body and mind, through trial and error and more formal education, acquires specific skills in perceiving and understanding environmental information. Sometimes this is quite conscious and intentional, as in the training of aesthetic appreciation. Most commonly it is a kind of cultural reflex, everyday, taken for granted and somewhat unconscious. We are often only aware of the group- or culture-specific nature of our perception when encountering people who do not share this style of perception (Hall 1969).

The priorities, or ways of perceiving, and meanings attached to perceptions vary widely between cultures and over time, and within cultures differences can be observed between different age groups, the sexes and socio-economic classes (Tuan 1974, 1993). Hall (1969) has explored the contrasting experience of Arab, Japanese and American perceptions of space and behaviour with respect to other people in space. Carpenter (1973), Lopez (1986) and many other anthropological investigators have shown how different cultures display different styles and depths of perception. Berger (1972, 1980) explored the socially specific nature of Western aesthetics and feminists have considered gender-specific issues in the socialisation of perception, both women's perception of themselves and others, male perceptions of women and the positioning of the female in culture (Pollock 1988; Rose 1986; Irigaray 1985a, b).

22

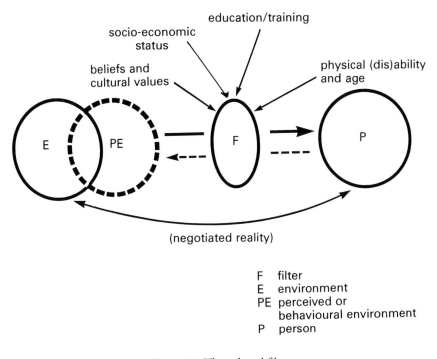

Figure 2.2 The cultural filter
Source: developed from Jeans (1974)

Jeans (1979) provides a useful summary model of the cultural dimension of human perception and the senses. Developing from the earlier behavioural environment ideas, he distinguishes between a 'real' and 'perceived' environment. It is the perceived environment which is important in decision-making. He models differences in styles of perception in terms of an intervening *cultural filter*. Figure 2.2 develops Jean's model by identifying specific elements which contribute to the cultural filter – shared values and beliefs, education and training, physical ability and age, and socio-economic status or class.

The cultural filter reflects the shared values and taken-for-granted practices of a specific society (or sub-group within a society). In other words, we see, hear, smell, taste and touch the world through the mediation, the filter or lens, of our social milieu, the context within which we have become socialised, educated and familiarised. Even within one society, there are various subsidiary filters associated with an individual's socio-economic status, education, age and gender. Thus, the cultural perspective reminds us that perception is more qualitatively variable and creative than mechanistic stimulus–response type models might suggest.

The cultural dimension can also be interpreted in interactive terms as the establishment of an accepted 'language' or game rules for person–environment and person–person relationships. Within this 'game' particular sense modes and/or modes of thinking may be given greater prominence and/or become associated with particular life situations. In Western cultures, for instance, the use of touch in interpersonal relationships varies quite markedly between Anglo-American and Latin peoples. The latter make much bodily contact even in everyday public interactions, whilst British and American people reserve interpersonal touching to more private contexts with close friends, family and sexual partners. It is also widely argued that in modern Western culture the eye has gained much dominance over the other sense organs, particularly the middle-class, male eye (Irigaray 1978). The hegemony of the visual in Western culture has perhaps been overemphasised at times, but it is interesting to observe that some of the most advanced technologies of our society are visually based, that the visual arts and mass media are major features of our culture, that many European languages have a particularly rich visual vocabulary which is frequently used metaphorically to describe other dimensions of sensuous experience and that, when asked which sense they would least like to lose, many people identify sight. The importance of the visual seems to extend also into patterns of thinking. The detached, objective view that has emerged as an ideal of science since the Renaissance perhaps has roots more in a visual than an olfactory or tactile sensibility. The notion that seeing is believing is quite a modern, culture-specific notion and, interestingly, doubting Thomas wanted to touch Jesus before he believed (Pocock 1981a).

In other cultures and times, the sensuous priorities and modes of thinking appear to differ from our own. The aboriginal eskimos, according to Carpenter, 'define space more by sound than sight' (1973: 33). Their world is of event rather than image, of dynamics and change rather than scenes and views. In considering such other cultures as these with our eyes, ears, noses and hands, socialised as we are into our own cultural frame of reference, all too easily we can misunderstand the nature of their experience and undervalue their understanding of the world around them. As Pollock writes: 'culture can be defined as those social practices whose prime aim is signification, i.e. the production of sense or making of orders of sense for the world in which we live' (1988: 202). Perception is an essential part of this cultural process of signification, it is culturally specific.

3

THE CHARACTER OF
SENSE

INTRODUCTION

The commonly recognised senses – touch, taste, smell, sight and hearing – may seem on reflection to be rather arbitrary, yet have endured as useful categories of sensuous experience. A number of researchers have offered definitions of each sense and suggested additional senses, most notably ones associated with the sensory properties of the body itself, the sense of balance and kinesthesis (or the perception of muscular effort or locomotion) (see Gold 1980: 50–51). However, paying too much attention to the identification of distinct senses can lead us to overlook the important inter-relationships between the senses and the multisensual nature of geographical experience.

Von Horbestel argued that 'one must search in order to find the private property of any one sense' (1927: 84). Whilst individual sense organs can be identified, the propensity of any one sense is often dependent on more than a single dedicated organ and is realised through a number of sensory faculties, some of which could be identified as of a different sense mode. This is especially the case with the intimate relationship between taste and smell (Gibson 1968), and is also evident in the overlap between tactile conscious-ness of vibrations and auditory experience. Sensuous experience is, in any case, often a complex of senses working together offering a range of 'clues' about the environment through which the body is passing. Yet, Engen echoes many other observers when he notes that 'each modality seems to do certain tasks well' (1982: 59), and therefore the distinctive role of individual senses is discernible. This ambiguity of distinctive senses and sensuous experience means a categorisation of the dimensions of sensuous experience must be interpreted with care and the value of such categories is primarily analytical.

Borrowing terms from McLuhan (1962), one might also argue that each sense – touch, smell, sight, hearing – is both a medium and a message. First, *a sense is a medium* through which 'information' about the environment is gathered. Each sense organ is receptive to particular types of environmental information – material surfaces, chemical compounds, ambient light, air vibrations, and second, *a sense is a kind of message*, or a distinct perspective on the world. Each sense organ by its very nature is selective of the environmental information it gathers, so filtering and structuring the information into particular messages.

This reminds us of the duality of the term 'sense' – sensation and meaning: medium equates with sensation, message with meaning. Thus the medium is the message and the message is the medium. The senses gather information but also contribute to the definition of that information, that is, participate in sense making. Sensuous geography, therefore, is an interaction with the environment both as given to the senses and as interpreted by the senses themselves in conjunction with the mind. As Tuan has recently written: 'the senses, under the aegis and direction of the mind, give us a world. Some are proximate; others are distant ...' (1993: 36). But what of the categories of sense?

CLASSIFICATION OF THE SENSES

There is a fundamental ambiguity in any study of the role of the senses – in self-awareness, human relationships, aesthetic appreciation and especially in understanding the role of the senses in environmental knowledge. The more one explores the nature of sensuous experience, phenomenologically or scientifically, the more one becomes aware of complex associations, substitutions, and transformations operating between the different sense organs, the human body itself, and the integral role of various complex mental processes. What might first appear to be a visual perception may on closer inspection be seen to include important auditory, olfactory and tactile components. Nevertheless, it does seem possible to identify different dimensions of sensuous experience akin to the usual classification of the senses which offer distinctive elements in a multisensual experience.

In the present study of sensuous geographies we investigate four senses, or *dimensions* – touch, smell, hearing and sight. Each of these seems to offer characteristic features to geographical experience. Traditionally, smell and touch are described as the intimate senses, associated with the body and what is in reach of the body, and sight and hearing are described as the distant senses, permitting the perception of a world beyond the body's immediate reach (Tuan 1993; Hall 1969). Whilst this distinction has much merit, it does tend to over-generalise. Closer examination reveals that each sense dimension operates over both intimate and distant ranges, though with different efficiency. Touch may not contact a distant movement in the environment directly, but feel the vibration generated in materials in contact with, or reach of, the body. Also, it is possible for the reach of any one of the senses to be extended through various tools invented by human ingenuity, such as spectacles, hearing aids and many scientific instruments which amplify sensual stimuli. To describe a sense as distant is, of course, not to neglect its important intimate role. Some things can only be seen or heard from close range. Furthermore, the distinction between intimate and distant senses is essentially a spatial classification and neglects important temporal characteristics of sensory experience. Of course, there is also a metaphorical interpretation of this distinction. It is sometimes assumed that touch and smell are more imtimate and emotional senses, whilst sight and hearing are deemed to be colder and more detached. This metaphorical interpretation of the intimate/distant distinction is problematic and is deliberately avoided here.

Visual experiences present themselves immediately as a scene, image or

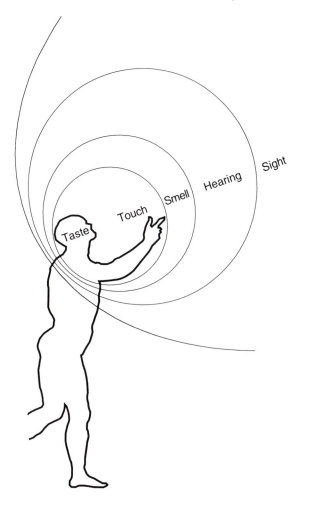

Figure 3.1 The range of the senses

Source: Skurnik and George (1967: 14). Copyright 1964, 1967 by Larry S. Skurnik and Frank George. Reprinted by permission of the authors

view, whilst auditory experiences unfold over time, like a tune, song or story. Smells equally have a complex temporal profile. Sensuous geography is in fact both spatial and temporal in character.

Any classification of the senses is first and foremost an analytical device, a simplification and an abstraction. It is important to consider the underlying assumptions on which the classification is based. Beyond the intimate/distant distinction, it is possible to classify the senses in a number of other ways. Tradition suggests five basic senses – touch, taste, smell, hearing and sight – yet experience discovers many ambiguities and additional sensory dimensions (Tuan 1974, 1993). Gold suggests that there are perhaps as many as ten basic senses. 'Besides taste, smell, sight, and hearing, there are four tactile or skin senses of pressure, pain, cold and warmth, and the two body senses of balance (the vestibular sense) and kinesthesis (the sense of movement in any part of the body)' (1980: 50). Each of these clearly has an impact on geographical experience.

However, one could quite easily offer a number of other classifications since – five or ten – they are to a large extent arbitrary, rooted in key assumptions (or presuppositions) and interpreted in culturally specific ways. This everyday classification with five or more senses assumes that each sense can be correlated with a specific sense organ which responds to a specific stimulus (or stimuli). It also implies that each sense is exclusively concerned with a particular aspect or dimension of geographical experience. Therefore, the eye is a specialised organ for seeing, it responds to light (photons), and the ear is a specialised organ for hearing, it responds to air vibrations. This assumption of exclusive sense-to-stimulus relationships is far too simplistic. It denies the complexity of the sense organs. For instance, the ear region is concerned with balance as well as hearing, and in any case it is widely understood that the body itself can feel vibrations both through the air and through the resonance of other substances and so 'hear' in a broader sense. Touch is far more than just the fingers, or a limited number of tactile receptors, but includes the whole skin surface and as such the skin becomes the largest sense organ (Montagu 1971). The tactile sensation is at the same time a sense of surface and form, texture and temperature, pressure and movement. Moreover, it is now widely accepted by scientists that a distinction between taste and smell is difficult to sustain and at the very least these senses always operate closely together (Gibson 1968). Of key importance to all the senses is the ability of the body and its parts to move, that is, its locomotion, and to manipulate and inspect things. We do not just sense passively but actively, sensuous experience is exploratory and this exploration marshals all the senses cooperatively.

Gibson (1968) suggests an alternative classification of the senses which seeks to be more precise and accurate with regard to actual everyday experience. He classifies the senses in terms of five perceptual systems. These are defined in terms of specific functions and relate to complex associations

Table 3.1 Perceptual systems

Name	Mode of attention	Receptive units	Anatomy of the organ	Activity of the organ	Stimuli available	External information obtained
The basic orienting system	General orientation	Mechano-receptors	Vestibular organs	Body equilibrium	Forces of gravity and acceleration	Direction of gravity, being pushed
The auditory system	Listening	Mechano-receptors	Cochlear organs with middle ear and auricle	Orienting to sounds	Vibration in the air	Nature and location of vibratory events
The haptic system	Touching	Mechano-receptors and possibly thermo-receptors	Skin (including attachments and openings) Joints (including ligaments) Muscles (including tendons)	Exploration of many kinds	Deformations of tissues Configuration of joints Stretching of muscle fibres	Contact with the earth Mechanical encounters Object shapes Material states, solidity or viscosity
The taste–smell system	Smelling	Chemo-receptors	Nasal cavity (nose)	Sniffing	Composition of the medium	Nature of volatile sources
	Tasting	Chemo- and mechano-receptors	Oral cavity (mouth)	Savoring	Composition of ingested objects	Nutritive and biochemical values
The visual system	Looking	Photo-receptors	Ocular mechanism (eyes, with intrinsic and extrinsic eye muscles, as related to the vestibular organs, the head and the whole body)	Accommodation Pupillary adjustment Fixation, convergence Exploration	The variables of structure in ambient light	Everything that can be specified by the variables of optical structure (information about objects, animals, motions, events, and places)

Source: Gibson (1968: 50)

of sense organs and muscles, and physical and mental processes. Nevertheless, Gibson maintains the distinction between basic bodily orientation and four further categories of environmental information – auditory, haptic (touch), taste–smell and visual. Gibson's (1968, 1974) perceptual theory is particularly relevant to geography since the primary work was on spatial perception (chiefly visual). His ecological theory of perception emphasises the role of the environment itself in structuring sense information and accommodates – in fact necessitates – the inclusion of a mobile observer in a dynamic world (see Chapter 2).

The perceptual systems approach treats the senses 'as active rather than passive, as systems rather than channels, and as related rather than mutually exclusive' (Gibson 1968: 47). Traditionally, it has been assumed that the sense organs provide the only sources of knowledge about the world and that they are each the channels for special qualities of experience. Gibson rejects this view as an oversimplification – perception is far broader and more variable. 'Sensation is not the prerequisite for perception and the sense impressions are not the raw data of perception – that is, they are not all that is given for perception' (Gibson 1968: 48). The human senses defined as 'active systems' are not passive receptors of sensations but are each actively engaged with the environment – exploring, collecting, responding to and participating in the evaluation of stimuli of many different kinds. Therefore, Gibson refers to the senses as exploratory faculties: listening, touching, smelling, tasting, and looking. Also, the distinction between 'motor' and 'sensory' organs is regarded as incorrect since perceptual systems involve their integration. Movement is of two types: exploratory and performatory, the first serves perception and the second behaviour. Furthermore, it is not possible to produce an exhaustive inventory of perceptions since the potential stimulus information is unlimited – some systems will also pick up the same information as others, redundant information, whilst some will not, and they will cooperate in varying combinations. The perceptual systems classification of the senses can accommodate both the distinctive activity of individual sense dimensions and the multisensual nature of human perception, integrating physical and mental processes, muscular and sensory components, and reception of stimuli and behavioural responses.

BODY AND SENSE

All sense perception is merely one outcome of the dependence of our existence upon bodily functioning. Thus if we wish to understand the relation of our personal experience to the activities of nature, the proper procedure is to examine the dependence of our personal experience on our personal bodies.

(Whitehead 1938, in Espeland 1984: 131).

30

The body is an essential part of sensuous experience: as a sense organ in itself (including the skin), as the site of all the other sense organs and the brain, and our primary tool for movement and exploration of the environment. Geographical experience is fundamentally mediated by the human body, it begins and ends with the body. This is the basic corporeality underlying all sensuous geography.

Corporeality subsists both in the structure of the body, as a physical object situated with other 'bodies' in a wider environment and with basic physical properties and needs, and the subjective experience of being a personal body, owned by an individual with a particular biography and situated in (and conditioned by) a given culture. Human bodies vary in their (dis)abilities with health and age, education and training, gender and socio-economic status of the owner. Whilst a physical entity, the body is also experienced as a culturally defined artefact. It has a history and is situated within specific cultural practices (Featherstone *et al.* 1991).

The human body is an important part of sensuous geography both in itself and in relation to the other senses. Four primary aspects of the body may be highlighted.

1   The body gives us an *orientation* in the world through its physical structure (the geometry of the body – front, back, left, right, up, down (see Tuan 1977) – and disposition of sense organs) and its own sensory capacities (such as its relation to the force of gravity) (Gold 1980).

2   The body gives us a *measure* of the world, that is forms a basic yardstick by which we appreciate space, distance and scale (through the size of the body with respect to other features of the environment and again its basic geometry) (see Tuan 1977).

3   The *locomotion* of the body and its parts offers us the potential to explore and evaluate our environment, and to change our location to satisfy our needs. The mobile body offers continual new perspectives on the world, allowing a richer and more subtle perception by all the senses.

4   The body as *coherent system* helps to integrate and coordinate (through its muscular structure and the functions of the brain) the sensuous experience generated by the various sense modes. It also gives a sense of wholeness and the relationship between parts.

The body contributes both to spatial and temporal perception, being like a ship and its anchor in our life-long geographical experience. It mediates between us and the environment, giving us access to a world beyond itself. In fact, without our bodies we would have no geography – orientation, measure, locomotion, coherence. It is therefore surprising that perception studies have largely focused upon the key senses – predominantly sight and hearing – and neglected the body as grounding of experience and as a sensuous geography in itself. Tuan (1977) has considered the geography arising out of the disposition of the body, that is the way in which our bodies give us orientation in and a measure of the world (see Figure 3.2). Hahn (1989) has discussed the body image in the geographical experience of the disabled, but otherwise geographers have tended to ignore the sociology and anthropology of the body and accepted taken-for-granted conceptions of the body.

Tuan has explored the actual and metaphorical relationships between the human body form and built forms and landscape conceptions in different cultural contexts (Tuan 1974, 1977). He recognises wide cultural variations in spatial concepts, but also identifies many cross-cultural similarities which 'ultimately rest on the fact that man [humankind] is measure of all things' (1977: 334). The human body form defines in a literal way our geographical experience and generates a somewhat unique geography due to our unusual upright posture. Tuan identifies two basic features of the human body as measure: the posture and structure of the body itself, and the nature of bodily relationships between human beings.

'The human being, by his presence, imposes a schema on space. Most of the time he is not aware of it' (Tuan 1977: 36). The posture and structure of the body generates a particular local geography – up and down, back and front, left and right – which are also culturally associated with meanings such as sacred and profane. We are the centre of our world, always experiencing the environment firstly from within this 'circumambient space' or immediate geography. As one moves, one's left and right, back and front, and so on also move. This immediate geography is extended by the body's senses, the intimate senses of touch and smell and the distant senses of sight and hearing. And more directly, the locomotion of the body allows it, with the aid of memory and expectation, to develop a wider 'map' of the environment through which it travels. Technology also extends the reach of the body and can give us a sense of experiencing a world apart from the body. Here, technologies such as the telephone and television are everyday examples.

Frequently, 'the body' has been adopted as a metaphor for wholeness and integration of geographical processes and the earth itself. The organism metaphor has a long and varied history in geographical thought (Buttimer 1993). Herbertson (1965) suggested that the earth is like an individual and that geographic regions equate to its organs, tissues and cells. Davis (1909) utilised the developmental dimension of the organism analogy to describe landscape change in terms of youth, maturity and old age. More recently, Porteous (1986a) has explored the body–landscape metaphor, but Stoddart (1967) is perhaps correct to see a certain immaturity in geographical use of an organism/body idea as analogy for geographical explanations and correctly labels it of 'dubious quality'.

Sociologists and anthropologists have recently taken a great deal of critical interest in the cultural definition of the body and its role in society and as a site of meaning (Turner 1984; O'Neill 1985; Featherstone *et al.* 1991; O'Brien

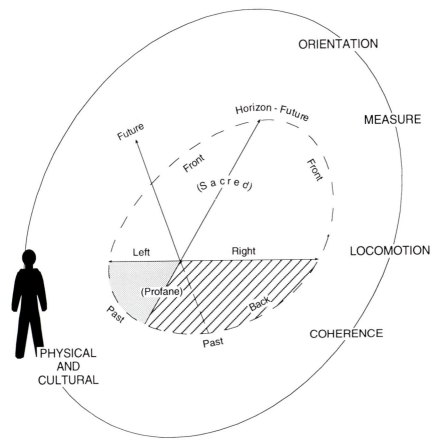

Figure 3.2 The geography of the body (space projected from the body is
biased toward the front and right. The future is ahead and 'up'.
The past is behind and 'below')
Source: developed from Tuan (1977: 35) with additions by present author

1989; Pollock 1988; Rose 1986; Bottomley 1979; Brian 1979; Foucault 1980;
Hargreaves 1987). There are many classifications of the history of the body.
In the European context, three key phases may be identified: the holy versus
the earthly body of the medieval period, the body as biological machine in
the following period and, in the twentieth century, the body as visual image
and commodity. Frank's (1991) typology of body styles offers a useful
framework within which to identify the different ways in which the physical
body is constituted culturally (Figure 3.3).

Frank argues that the body is constituted in the intersection of an
equilateral triangle the points of which are institutions, discourses and
corporeality. Discourses are 'cognitive mappings' of the body's possibilities

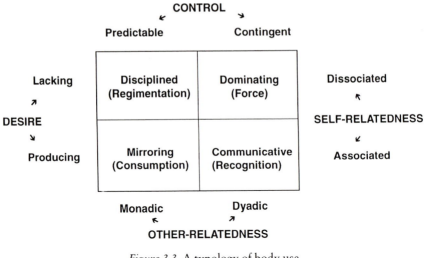

Figure 3.3 A typology of body use
Source: Frank (1991: 54)

and limitations which bodies experience as already there for their self-understanding. These exist in on-going practices. Institutions, on the other hand, are specific in time and space. They are structures, with a physical place where one can go at a particular point in history. Both these elements are built and decay as tangible geographies of the body. However, bodies do not emerge out of discourses and institutions, but are only modified by them. Bodies emerge out of other bodies – specifically women's bodies (O'Brien 1989). This is the corporeality of bodies. Bodies exist with space and time, a physiology and a biography. The experienced body is continually constituted and reconstituted through this complex of corporeality, institutions and discourses.

The four sides of matrix represent four 'action questions':

1 *Control* – the body must ask itself how predictable its performance must be.
2 *Desire* – the body is concerned with whether it is lacking or producing.
3 *Relation* – the body must have some sense of its relation to others. Is it monadic and closed in on itself, or dyadic and existing in relation of mutual constitution with others?
4 *Self-relatedness* – does the body consciousness associate itself with its own being, particularly its surface, or dissociate itself from that corporeality?

These four questions generate the matrix of four cells representing body uses, or styles: disciplined body, mirrored body, dominating body and communicative body. These are ideal types and actual bodily experience will only

34

approximate these. In fact, Frank notes that bodies do not stay long in any one use but continually change (Frank 1991: 53). Each body use resolves specific problem areas. The disciplined body responds to a problem of control, making itself predictable and regimented. The mirroring body remains predictable through its pattern of consumption. Both body uses are monadic, but whilst the disciplined body moulds itself in reference to a set of social or personal values or standards, the mirroring body constitutes its objectives in its own self-reflection. Consumption is not so much 'use' as the endless assimilation of the world's objects to one's body. In terms of sense and the body, the disciplined body seeks to train itself and to utilise its muscles, mind and senses to the full potential. The mirroring body absorbs all the sensuous experiences offered to it, accepting them and even desiring them, wanting to contain all that can be sensuously experienced within the body's experience.

The dominating body is almost exclusively male. It is characterised by a sense of lack and fear. It is dyadic and thus it seeks to dominate others in order to overcome its own lack, unlike the disciplined body which dominates itself. Dominating bodies need to seek out other bodies as sub-human enemies which they can fight and maim. Dominating bodies seem to enjoy extreme sensuous experiences of the suffering of others. The communicative body is 'less reality than praxis' (Frank 1991: 79). The other body uses exist, whilst the communicative body is more an ideal. It is a body in the process of creating itself. It is dyadic, and so its relation to itself and to others is not a mirroring but a self-realisation. In this sense, this body enacts a kind of phenomenological reflection on its sensuous experience and constitution as a body relative to a world, a body-in-the-world.

The body is therefore both a physical entity with an immediate geography and culturally defined in terms of a style or body use. Thus the body is constituted in a similar way to 'sense(s)' – as sensation and as meaning.

GENERAL PROPERTIES OF THE SENSES

In the following chapters we will focus on the distinctive contribution of each sense – touch, smell, hearing, sight – to geographical experience. However, it is important to remember that this is an abstraction. Whilst it is possible to identify the apparent dominance of a specific sense in a given situation, on closer analysis all geographical experiences are made up of a complex of sensuous information combining activities of the sense organs, the body and its limbs, and mental processes (memory and expectation, analysis and evaluation). Nevertheless, it is useful to pick out the distinctive contribution of each sense in geographical experience, remembering that actual geographical experience is a complex of sensuous (and emotional) geographies. The emotional response to the environment is a closely related topic but one of which space does not permit closer analysis here (see Tuan 1979b).

Describing and also understanding the role of the senses in geographical experience is complicated by the limitations of language. In English, for instance, visual, tactile and auditory dimensions are relatively well served by a rich vocabulary, but smell and taste are constrained by a limited specific vocabulary. Metaphor plays a key role in descriptions of sensuous experience, and sometimes these are so everyday that we forget that they are metaphors. The use of metaphors to describe sensuous experience also reflects the basic multisensual character of geographical experience and the complex inter-relationship of the senses. We may refer to the tactile properties of a picture seen by the eye, the resonant qualities of a sharp taste, and the visibility of a smell. In exploring sensuous geographies we must be critically aware of the language we employ and presuppositions it contains.

Though in Part 2 we will consider each sense in turn, certain common characteristics can be noted. The various senses relate and combine in complex ways to generate specific geographical experiences. Five key characteristics are identifiable in touch, smell, hearing and sight, individually and in their inter-relationship.

First, *cooperation* – the senses may operate together in many possible combinations and the sum of their partnership exceeds the effectiveness of each alone in discerning environmental details. A positive combination of the senses can enhance perceptual accuracy, but on occasion the senses may give conflicting information and so undermine clarity of environmental perception. Sometimes, one sense may effectively substitute for another, acting as a surrogate and provide adequate comparable environmental information. The experience of the blind and deaf, for instance, illustrates this well (e.g. Hull 1990; Wright 1990).

Second, *hierarchies* – although in general all geographical experience is multisensual, not all the senses play an equal role and one can identify hierarchies of sense, actual or only apparent. A certain sense may appear to play a dominant role in characterising a specific experience and other senses may appear to be subservient. The dominant sense may take a lead in initiating an experience or in establishing a general framework for geographical understanding which is followed through, filled out and clarified by the other senses. This hierarchy will differ from situation to situation, such as between object exploration and space perception, and day-time and night-time spatial orientation.

Third, *sequences* – the senses may appear to take on a particular order in geographical experience without any implication that one sense is more important than any of the others. Likewise the different senses each contribute information for the overall geographical understanding of a specific situation. Montagu (1971: 236) has suggested that in the biological development of humans the senses 'develop in a definite sequence, as (1)

36

tactile, (2) auditory, and (3) visual. As the child approaches adolescence the order of precedence becomes reversed, as (1) visual, (2) auditory, and (3) tactile', at least in North American culture.

Fourth, *thresholds* – the different senses may come into operation at specific levels of stimulation, or within the context of certain excitations of the other senses. These threshold relationships may be as much culturally learnt as biologically determined. Novelty may lower a sensuous threshold in some situations, whilst familiarity may raise the threshold of response. Thus, we may notice quite subtle sensuous details in places we visit for aesthetic pleasure, as when we walk in the countryside or go on holiday to foreign countries, but by contrast we may be largely oblivious to quite strong sense stimuli (smells, sounds, etc.) in our local town or our own home which a visitor, in turn, clearly notes. Habituation is an important dimension of sensuous experience.

Fifth, *reciprocity* – it has been argued that each of the senses also establishes a relationship between the sentient and the environment and things in that environment. At its most attentive, each sense can imply a kind of reciprosity. This is most evident in the case of touch, where to touch something also implies permitting that something to touch you (Montagu 1971). Less directly, coming close enough to identify the smell of someone also permits them to smell you. For the distant senses, this reciprocity is more indirect and sometimes is broken. To get a good view can mean putting oneself in the position of being seen, but as in Appleton's prospect–refuge theory (1975) one can see without being seen. Likewise it is possible to hear without being heard. Therefore, reciprocity is a potential of each sense but not a necessary characteristic of all sensuous experience.

SENSE AS GEOGRAPHY

The senses are geographical in that they contribute to orientation in space, an awareness of spatial relationships and an appreciation of the specific qualities of different places, both currently experienced and removed in time. The senses offer important media through which space and time is experienced and understood (made sense of). Yet also, each sense mode seems to offer its own distinctive character to that experience and, in particular contexts, a certain sense and a specific style of operation of that sense (which is biologically and culturally determined) may play a hegemonic role in establishing geographical meaning. Thus the multisensual nature of geographical experience is not even but variable across space and through time, between individuals and communities, between cultures and periods. Sensuous geographies are various in character and changing, and any general characteristics recognised are specific to a given socio-historical context.

In Part 2, each sense chapter begins with initial reflections on the characteristic geography of the specific sense mode. The sensuous matrix

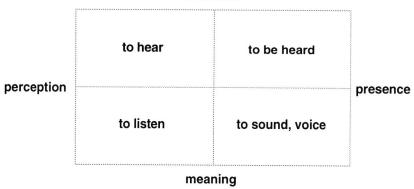

Figure 3.4 Example of a sensuous matrix (e.g. auditory experience)

identifies the main dimensions of sensuous geography (Figure 3.4). The vertical axis represents the integration of physical and mental processes, that is sense as sensation and meaning. The horizontal axis represents the two facets of geographical reciprocity: the 'gathering' of a world around us (passive and active perception) and the projection of the individual's own presence into space, creating particular sensuous spaces (unintentional side-effects of being present and intention to communicate a presence to the world or establish a claim to space).

The following chapters explore this sensuous geography.

Part II
SENSE, SPACE, PLACE

4

HAPTIC GEOGRAPHIES

INTRODUCTION: THE HAPTIC SENSE

Touch geographies are the sensuous geographies arising out of the tactile receptivity of the body, specifically the skin, and are closely linked to the ability of the body to move through the environment and pick up and manipulate objects. Touch can be both passive and active, a juxtaposition of body and world and a careful exploration of the size, shape, weight, texture and temperature of features in the environment. Touch is above all the most intimate sense, limited by the reach of the body, and it is the most reciprocal of the senses, for to touch is always to be touched (Montagu 1971).

There is a rich array of metaphors which attest to the importance of touch in everyday social and geographical experience. We refer to 'keeping in touch' by letter or telephone, describe people as being a 'soft touch', talk about 'rubbing someone up the wrong way'; we may try to avoid 'touchy' subjects in conversation, or describe a kind gesture as 'touching'; we talk of 'smooth transitions', the first draft as a 'rough copy' and the rogue as a 'rough diamond'. Many different emotions can be associated with touch – from caring and love to disgust and hate. It is therefore a highly significant dimension of the human experience, both in person–person and person–environment relationships. We might lose any of one or more of the other senses – sight, hearing, smell, for instance – but to lose an ability to feel, that is, touch, is to lose all sense of being in a world, and fundamentally of being at all (Tuan 1993). If we lose the skin, the primary haptic layer, we cannot survive long (Montagu 1971). To be skinned alive is perhaps the greatest horror short of prolonged and agonising death, since the body loses its haptic layer with a massive tactile information overload, a final assertion of its all-encompassing importance.

The label 'haptic' is adopted for two reasons. It avoids the superficial connotations associated with the everyday word 'touch', and in particular the assumption that touch geographies are only the sensuous experiences of the fingers. Montagu (1971) amply shows how tactile experience involves the whole body, and that touch is not restricted to the fingers but is a property of the whole skin covering the body. 'Haptic' also refers to Gibson's 'haptic system', a functional definition of touch as a system involving the coordination of receptor cells and the muscles of the body (Gibson 1968). Gibson suggests that touch refers to two distinct faculties. It is pressure on the skin, or literally contact between the body and its environment, and it can also refer to kinesthesis, that is the ability of the body to perceive its own motion. Touch is, therefore, about both an awareness of presence and of locomotion. Together these can be described as the haptic sense (from the Greek, 'to lay hold of').

> The haptic system ... is an apparatus by which the individual gets information about both the environment and his body. He feels an object relative to the body and the body relative to an object. It is a perceptual system by which animals and men (humans) are literally in touch with the environment.
>
> (Gibson 1968: 97)

In this chapter, 'haptic geographies' refers to touch as an active sense which is integrally involved with the locomotive ability of the body and specifically focuses upon the role of touch in the perception of space and relationships to place.

In Chapter 3, touch was distinguished from the body senses (of balance, gravity and movement). Here, 'haptic' refers to the tactile receptivity of the skin, the movement of the body parts and the locomotion of the whole body through the environment. To this extent, kinesthesis is included in the label 'haptic'.

Haptic geographies are often overlooked, since the tactile experience is such a continuous and taken-for-granted part of everyday encounter with the environment. This chapter falls into five parts: an explication of the general character of haptic geographies, reflections on the variety of individual touch geographies, an exploration of the cultural differences in touch behaviour and, finally, some reflections on the haptic dimension of place experience in contemporary Western societies and the role of changing technologies and cultural values. Throughout, emphasis is placed upon the distinctiveness of touch as a dimension of geographical experience, but it must always be remembered that the full significance of haptic experience lies within the wider context of multisensual experience of the environment. Interestingly, touch is sometimes seen as 'containing' all the other senses, or as a foundation upon which the subsequent sensuous experience is built (see Montagu 1971: 270). Therefore, haptic geographies might be seen as both part of a wider multisensual geography and a kind of foundation for such a geography.

TOUCH, SPACE AND PLACE

The skin mediates between the body and the surrounding environment. The surface area of the skin has an enormous number of sensory receptors receiving stimuli of heat and cold, pleasure and pain, and an infinite range of tactile sensations. It is estimated that there are some 50 receptors per 100 square millimetres. Tactile points vary from 7 to 135 per square centimetre. The number of sensory fibres from the skin entering the spinal cord by the posterior roots is well over half a million (Montagu 1971: 3). The skin is by far the largest single organ system of the body and therefore its importance to geographical experience cannot be overestimated. (Experientially, there are many inter-relationships and overlaps between the different senses, none more so than with touch as a ground base experience. One can also identify structural similarities: e.g. the retina of the eye and the membranes in the ear are highly specialised kinds of 'skin'.) The relative neglect of haptic geography is, therefore, a major omission by geographers (see Pocock 1993; Tuan 1993).

The sense of touch is the earliest to develop in the human embryo. When it is

> less than one inch long, less than eight weeks old, light stroking of the upper lip or wings of the nose will cause bending of the neck and trunk away from the source of stimulation. At this stage in its development the embryo has neither eyes nor ears. Yet its skin is already highly developed, although in a manner not at all comparable to the development it is still to undergo.
>
> (Montagu 1971: 1–2)

At birth the skin has to make significant adaptations as the baby moves from an aquatic womb environment to a relatively dry air environment. Therefore, not only is the skin the first means by which the organism acquires a geography – a sense of the world – but it is also, already, right from the start, a highly adaptable perceptual system.

So often we take for granted this vital tool of the haptic system, the skin and its receptors. Yet, it is both a highly versatile organ and a very delicate one. The skin can be easily damaged but, unlike the other sense organs, it is capable of relatively rapid regeneration throughout most of the life of the organism, except in the worst cases of damage. Physiologically it fulfils four key functions: protecting the underlying parts of the body from mechanical and radiation injuries and invasion by foreign substances and organisms; as a sensory system it provides a wide range of information about the surrounding environment; it regulates the temperature of the body and is involved in the metabolism of fats and, through perspiration, of water and salt. The skin is a remarkably sensitive interface between our body and the environment and gives us a vast amount of vital information.

> Most of us take our skin entirely for granted, except when it burns and peels, or breaks out in pimples, or perspires unpleasantly. When we think of it at other times, it is with a vague wonder at so neat and

efficient a covering for our insides: waterproof, dustproof, and miraculously – until we grow old – always the right size.

(Montagu 1971: 5)

As the central element of the haptic system, the skin combines with the muscles of the body, the mobility of the body and its size and proportions as an important reference point, to permit us to explore the tactile world and feel the intricate details of objects in that world.

This haptic geography includes information about the size, weight and form of objects and their movement relative to the body (especially if relatively slow), the size, shape and depth of volumes, and myriad details about the various textures, temperatures, flexibility (or squashiness) and continuity of surfaces The haptic system, in conjunction with the brain, collates all this information to give coherent tactile sensations of recognised features in the environment.

Touch literally concerns contact between person and world. It is participation, passive and active, and not mere juxtaposition. The haptic system gives us the ability to discriminate key characteristics of the environment and our place as a separate entity in that environment or world, but it is not just a physical relationship, it is also an emotional bond between ourself and our world. Touch is a kind of communication between person and world, a corporeal situation rather than a cognitive positioning. The gentle touch is always more effective than mere words. Touch is direct and intimate, and perhaps the most truthful sense. 'Touch is the sense least susceptible to deception and hence the one in which we tend to put the most trust' (Tuan 1993: 45). Doubting Thomas had to touch Jesus to believe in the resurrection, the Pope kisses – touches with his lips – the ground of each country he visits. Touch, or the haptic system, provides both information about objects in the environment and their environment, and – by definition – our relationship to them and our participation in that environment.

The haptic experience is also one of both stationary objects and movement within and across space. Through touch we not only have access to the material world, but also a living one. Through the haptic system we are able to discern the living from the dead, the friend from the foe and, above all, ourselves within the context of an immediate world. Touch gives us a place in a world. To deny all touch – to lose touch – is to lose a world and, in effect, our sense of identity, even awareness of being (Tuan 1974). Hull in a 'diary' of his journey into blindness, a coming to terms with a world without faces and visual images, evocatively gives it the title: *Touching the Rock* (1990) with both tactile and religious connotations.

Touch is more than the action of the fingers feeling the texture of surfaces. Touch involves the whole body reaching out to the things constituting the environment and those things, or that environment, coming into contact with the body (Boring 1942). This is the basic reciprocity of the haptic system; to

44

touch is always to be touched (Montagu 1971) – though the intention may be the preserve of only one party. Reciprocity is, in this sense, often asymmetrical and typically so in contacts with inanimate materials. Three levels of reciprocity in environmental experience can be identified.

1   *Simple contact* – the juxtaposition of two surfaces against one another, often temporary, and often involving only one tactile sensitive agent. Such contact is generally passive touching or rubbing against or simple co-presence. An organism moving across an inanimate surface, sand or rock, or moving through vegetation (where the plants may 'feel' the contact but do not necessarily make that known to us) by definition is in simple contact with the environment but may not consciously take much notice of it.

2   *Exploratory activity* – an agent actively investigates the environment (organic and non-organic) but the environment explored is largely passive and does not respond to the touch nor appear to register its own tactile sensation. This is quite clearly a kind of asymmetrical reciprocity. The active agent is conscious of a stream of haptic sensations giving a rich supply of information about the environment explored.

3   *Communication* – the contact is actively intended, by one party or both, but each party responds specifically to the other's tactile stimulation and messages are exchanged. Stroking the cat, cuddling a friend and some encounters with certain plants, such as fly catchers, perhaps. This communication is generally a relationship between organisms and forms an important part of the bonding of community and the sense of establishing roots in a place.

Everyday haptic experience, of course, involves constant changes between these different levels, or types, of co-relation. Geographical experience arises out of all three. Exploration provides the fine detail of a complex tactile world, communication establishes a participation or belonging to that world (and in a sense establishing our place in it) and simple contact continually maintains, often subconsciously, a connectivity with the physical world and our own bodies. Exploration essentially distinguishes between different features, or objects or surfaces, in the environment and 'maps' spatial relationships. Communication establishes an active relationship to the world, transforming abstract spaces into meaningful places and perhaps – through the social dimension of touching – giving us roots in a place. Simple contact is geographically a simple confirmation that the world still exists and that we are 'with our feet on the ground'. If the taken-for-granted does not fit the reality, this may spark a shift to exploratory activity (as when a new stimulation appears).

Implicit in this abstract typology is a fourth level, sometimes experienced by people suffering certain mental conditions. In *The Betrayal of the Body*, Alexander Lowen (quoted by Montagu 1971: 205) describes the relationship of schizophrenia to the loss of early tactile experience. He shows how feelings of identity arise from feelings of contact with one's own body. To know *who* one is, and ultimately to know *where* one is, is grounded in contact with one's own body and, specifically, feeling an ownership of a body, one's own body in itself and relative to other objects in the environment. In schizophrenia the individual loses a sense of contact with a 'real' world and often feels strangely apart from the world. In extreme cases, the individual loses all sense of themselves and cannot distinguish between self and non-self. The extreme schizophrenic, who loses body contact completely and literally becomes out of touch with reality – that is image and reality become dissociated in the schizoid experience.

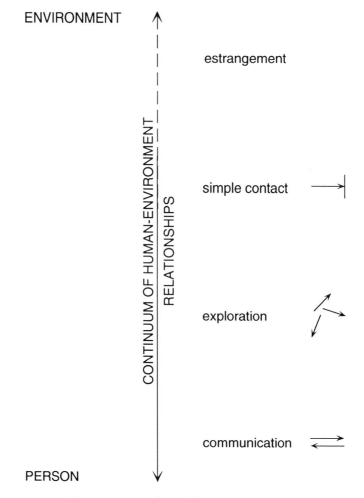

Figure 4.1 A typology of haptic relationships

46

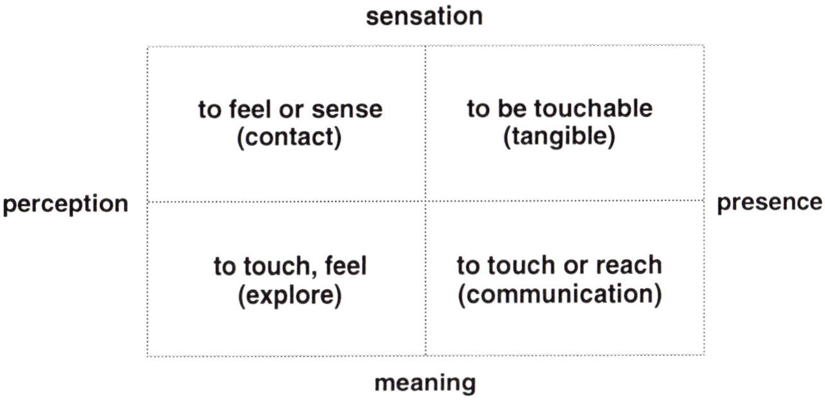

sensation

to feel or sense **(contact)**	**to be touchable** **(tangible)**
to touch, feel **(explore)**	**to touch or reach** **(communication)**

perception   **presence**

meaning

Figure 4.2 The haptic matrix

Closer reflection on geographical experience suggests a number of other finer divisions between these levels and reminds us that actual experience is conceptually a continuous movement up and down a continuum of haptic relationships between person and environment (Figure 4.1).

Touch is both active and passive. The ultimate reference for touch is the human body – its size and orientation, its own relative motion, temperature and humidity, as well as its tactile surfaces or haptic capabilities. Haptic geographies begin and end with the living body, literally and metaphorically. The body defines touch, or haptic, geographies both directly in its immediate exploration of a world and indirectly through extensions or tools, such as the walking stick, employed by individuals to extend the reach of the body and/ or amplify specific sensuous experiences.

The haptic sense is a complex of different kinds of relationship with the environment. We have already noted degrees of encounter (Figure 4.1). Seamon (1979) suggests a continuum of modes of environment–person encounter from full involvement to total detachment, which is similar to this typology. The haptic matrix suggests an alternative classification (Figure 4.2). Just as we might distinguish between seeing and looking, or hearing and listening, one can distinguish between feeling in general and touching (or feeling) in the specific senses. This is really a distinction between the passive and the active. Likewise, just as one is seen or heard, and presents a look or gives out a sound or gives voice, so one is in touch or contact with the environment as a physical presence and also one presents oneself as touchable or in reach. These two axes, sensation–meaning (sense) and perception–presence (relation), represent the overall structure of haptic geographies, not just about touching but being touched, of reaching out and making oneself within reach.

DIMENSIONS OF TOUCH

In the previous section we considered the general qualities of haptic experience. Reflection on our own experience, the sensuous geographies of children – sometimes noted for their greater use of touch – and the disabled, enables us to discern further details about the nature of haptic geographies. In the everyday experience of the able-bodied adult, the haptic dimension to geographical perception is often overlaid by visual and auditory information and thus tends to be overlooked, except when something is, for instance, particularly rough or smooth, hard or soft, hot or cold. The wearing of clothes, and especially gloves, can also mask or reduce tactile sensations. However, touch is an important component of geographical experience.

The haptic experience – a combination of tactile and locomotive properties – provides important information about the character of objects, surfaces and whole environments as well as our own bodies. Personal reflection on our own experience can identify a broad range of detailed information we derive every day from our use of the haptic system.

1 *surface* – texture, that is roughness or smoothness, and details of surface variation;
2 *geometry* – shape, dimensions or size, proportion and arrangement, generally relative to our human scale;
3 *material* – the mass or weight of objects supported by the body or parts of it, and the perception of material rigidity or plasticity;
4 *location* – distance from us, either in terms of direct or extended reach, and direction relative to our body orientation – front, back, right, left, up, down;
5 *energy* – the judgement of a wide range of temperatures of both objects and environments, and their dampness or dryness (relative humidity);
6 *dynamic* – vibration and locomotion, that is perception of movements 'within' objects and of objects though space, relative to our own body and/or other objects in reach.

Each of these sets of variables enables us to identify particular features of the environment and experience a geography of spaces and places of distinct character. Contained within each of these is relationship to the body and an experience of contrasts between different features in the environment. 'Like all the other senses, the tactile sense is activated by contrast – alterations of heat and cold, roughness and smoothness, lightness and weight' (Tuan 1993: 44).

Starting with the human body, one can recognise four kinds of touch which each give access to this detailed environmental information – global touch, reach-touch, extended touch and imagined touch (Figure 4.3).

Global touch represents the body's general contact with the environment. Global touch is the presence of the body in a context, a sense of itself within a world – this is related to Merleau-Ponty's body-subject concept (1962; also Espeland 1984). This can be enhanced by the movement of the body through space, across surfaces, through the air and water, or even by careful meditation on the body itself (see pp. 19–22). This is generally a passive experience, a general feeling of one's body and its intimate environment. This geography is not so much about distances and spatial relationships, as about a general presence such as feelings of uprightness or basic body orientation (especially with respect to gravity), temperature, humidity and perhaps the relative crowding or space in its most general sense. Of course, this generalised touch can be amplified by the other touch dimensions and reinforced by the other senses, notably sight.

The experience of global touch is quite difficult to describe. The blind frequently refer to a strong sense of the presence of objects not directly touched but nevertheless felt as a kind of pressure in the air around the body. This is sometimes called 'facial vision'. It illustrates the subtle nature of global touch and the multisensual reality of sensuous geographies. Hull (1990: 20) describes facial vision in tactile terms:

> the experience itself is quite extraordinary, and I cannot compare it with anything else I have ever known. It is like a sense of physical pressure. One wants to put up a hand to protect oneself, so intense is this awareness. One shrinks from whatever it is. It seems to be characterised by a certain stillness in the atmosphere. Where one should perceive the movement of the air and a certain openness, somehow one becomes aware of a stillness, an intensity instead of an emptiness, a sense of vague solidity. The exact source of the sensation is difficult to locate. It seems to be the head, yet often seems to extend to the shoulders and even the arms. Awareness is greater when the environment is less polluted by sound, and in the silence of my late evening walk home, I am most intensely aware of it.

Hull describes this as an intense experience of the presence of large objects before they are reached or one can touch them. He describes how he has gradually extended his perception of this presence of objects to include trees, lamp posts, walls and parked cars up to 6–8 feet away.

The experience of this kind of global touch is partly the result of specific environmental conditions, such as the quietness of an evening walk, and a developed attentiveness to body sensitivity often evident in the experience of the blind (see Lusseryan 1963, 1973; Hill 1985). Talking with a friend whilst on such an evening walk distracts one's attentiveness and the blind person can unwittingly walk into a lamp post (Hull 1990). This collision is not so much the result of blindness, as a sighted person might assume, but rather inattentiveness to alternative environmental information. Hull also observes that facial vision appears to disappear in busy city streets where the noise of traffic and the mass of cars and people blanks out such details, and in the home where carpets, curtains and furnishings probably reduce the clarity of echoes.

Whilst facial vision can be described as a kind of global touch, it could also be described as a subtle form of auditory perception. Hull describes it both as a kind of pressure on the skin and like small echoes set up between the body and other objects in the environment. The ambiguity of sense definition lies in part in attempts to describe kinds of experience not adequately assigned to existing sense categories. The body moving past large stationary objects or large objects moving past the body will generate subtle changes in air pressure which could account for a pressure change sensation. Equally, the same conditions, clarified by a quietness and stillness (such as on an evening walk), could also make evident minute echoes which might permit distinctions between open space and enclosed space, or the presence of large objects. Since auditory experience is grounded in receptivity to vibrations transmitted through air and the skin is sensitive to pressure (and vibrations through solid substances), air vibrations which can excite the membrane in the ear can also, it seems, excite receptors in the skin. In this sense the ear 'touches' like the skin touches the environment, but we discern the former as sounds and the latter as pressure, texture and so on. Facial vision, therefore, demonstrates the difficulty in accurately describing sensuous experience and assigning a specific sense label. The geography generated by this sense of presence is one of large objects relative to the body, more an early warning system than a map of one's surroundings. 'I must stop when I sense something, but not sensing something does not mean I can go ahead' (Hull 1990: 19).

In contrast to global touch, one can also identify a reach-touch. This is the form of touch we most immediately think of in everyday use of the term. It is the touch of the hands and arms, fingers and toes. It is exploring touch which reaches out to, takes hold of or feels the characteristics of objects and their relationship in the environment around us. This is active and generally grounded in intention. Utilising some of the most sensitive parts of the body, reach-touch is particularly discriminating of the characteristics of objects and the environment around us. This is the touch which is developed to a high level of sensitivity in the sculptor and frequently demonstrated by the blind. It enables us to discern characteristics, noted earlier, of surface, geometry, material, location, energy and dynamic. The limit of reach-touch is length (or reach) of the limbs used – the fingers, hand and arm, for instance. Individuals differ in their reach – most notably the child versus the adult – and therefore the extent of their haptic geographies. Movement of the arms, hands and fingers, or of the whole body through the environment (as part of passive or active touch) allows greater tactile detail to be gained and a more coherent sense of the world.

Reach-touch is employed both in exploratory activity and in supporting navigation through the environment. This is especially evident in both the young child and the haptic experience of the blind. Tactile stimulation plays an important role in the development of the young child (Montagu 1971). Information about the world derived from direct contact between the skin and the environment, both passive global touch and active reach-touch exploratory activity is important in establishing relationships with the mother and other interpersonal relationships and a relation to one's own body, in establishing a sense of one's place in a world as well as information about that world. Touch, therefore, is important in establishing trust and belonging, coordination and knowledge. Montagu (1971) presents a comparative cross-cultural analysis of child rearing practices and concludes that the North American child experiences a deprivation of tactile stimulation when compared to children in many other cultures. The American mother clothes her child and touches the child only for limited specific activities in caring for its needs. The child is carried in a pushchair rather than the arms or on the back, rocked in a pram or cradle rather than cuddled extensively, and discouraged from 'pawing' mother with its exploratory fingers. Margaret Mead has pointed out that the attention of the American baby is directed away from the personal relationship with the mother by toys which are introduced into the bath tub. Thus the baby's attention is focused upon things rather than persons (see Montagu 1971: 252–253). Montagu perhaps overemphasises this tactile deprivation, but it is clear that mothers in other cultures engage in more frequent and interactive tactile interaction with their young.

The young child explores a tactile world, beginning with its own body and the mother, and then going on to reach out to a wider world. This tactility is integrated with other sensory media. The thing touched is also sucked and tasted, or sniffed and smelled, or banged and heard, or turned around and viewed. This tactility is also closely allied to the movement of the body parts (arms and legs, hands and fingers, etc.) and the locomotion of the whole body through an environment. The child both explores and evaluates the environment with touch (and taste, etc.) but also navigates through that environment learning to support itself on specific objects as it progresses from crawling to walking. The child learns to discern the myriad tactile properties of its surroundings – surface textures, the solidity of objects, relative size and form, and moving through space using a kind of haptic map of the most familiar places.

Hull (1990: 28), in his reflections on blindness, notes that 'it is not easy for the sighted person to realise the implications of the fact that the blind person's perception of the world, sound apart, is confined to the reach of the body, and to an extension of his body which he can set up, such as a cane'. The blind thus illustrate both reach-touch and extended touch, that is, touch directly through the fingers supported by the reach of the arms, and indirect tactile sensation through tools of extension such as a cane. Hull (1990: 133) argues that touch gives the 'sense of real knowledge' and for the blind is also a source of great pleasure. Describing the experience of holding a model owl, Hull writes:

> He put into my hands a little stone owl about five inches high. It was squat and beautifully rough. The weight of it in my hand was satisfying. There was a carved, wooden owl from Africa. I admired the simplicity of the details, the warmth and smoothness of the wood, the way that the whole object could be contained within the hand. I am developing the art of gazing with my hands.

Hull summarises the alternative sensuous geography which he has come to enjoy and rely upon with the development of his blindness: 'weight, texture and shape, temperature and the sounds things make, these are what I look for now' (1990: 133). Four of these are, of course, haptic experiences.

> As long as the blind person has one hand free, he can see with it ... the free hand he uses to guide himself; he feels blind when he has no free hand; he feels blindfolded if both hands are occupied.
>
> (Hull 1990: 82)

Hence, it is difficult for a blind person to carry a tray full of glasses with both hands confidently, even in a familiar environment, since at least one hand (perhaps assisted by a cane) provides important tactile information. In this sense, one 'sees with the fingers' (1990: 82–83). Touch can therefore function as the eye does in giving navigational cues.

However, reach-touch is not the same for the blind person and the sighted person. A blindfold gives a false sense of what it is like to rely upon haptic experience, since without the possibility of sight at all – especially if one is born blind – touch is more than the mere feeling of things but becomes a highly tuned sense in its own right. Reach-touch is more than just the exploration of objects held in the hand, it can involve the whole body actively 'sizing up' and 'interacting' with tactile space. In getting to know the main altar of Iona Abbey, Hull describes how he used his fingers to explore it, stretched his arms over it, and pushed himself on to it with his feet hanging out over the front of the altar and reaching the back with his arms. Doing this physical exploration of the altar again and again, he was 'measuring it with his body'.

> It was bigger than me and much older. There were several places on the polished surface which were marked with long, rather irregular indentations, not cracks, but imperfections of some kind. Could it have been dropped? These marks felt like the result of impact. The contrast between the rough depressions and the huge polished areas was extraordinary. Here was the work of people, grinding this thing, smoothing it to an almost greasy, slightly dusty finish which went

slippery when I licked it. Here were these abrasions, something more primitive, the naked heart of the rock.

(Hull 1990: 163).

This is the richness of haptic experience; the mystery of tactile encounter. Here, we have a reach-touch from the lick of the tongue to the stretching out of the body as a measure.

The blind measure spaces, consciously and unconsciously, by the number of paces or strides walked across or along, up or down, and the memory of changes of direction. They also note the location of walls and fences used to confirm a location or direction at different points on a route. In familiar environments, this tactile mental map becomes imprinted in the mind and greatly assists everyday navigation. This is especially evident in the descriptions of their homes by both the blind and partially sighted people. The feeling of threat outside the familiar confines of home spaces directly reflects this lack of established tactile knowledge (see Hill 1985).

Figure 4.3 also indicates two other dimensions of touch which are especially developed in humans and offer an even wider geographical experience. Extended touch and imagined touch indicate that touch need not be an intimate sense in contrast to so-called distant senses, since in terms of spatial experience touch can reach far beyond the immediate geography of the body with the aid of tools and the imagination. Extended reach is touch mediated by or enhanced with technology. The white cane used by the blind person is only one example of numerous instruments used to extend the human reach. Often these instruments do not merely transmit contact information from a greater distance than our bodily reach but transform tactile information into other sensuous forms. The white cane not only offers vibrations as it is tapped, but also sounds which can be heard and used in

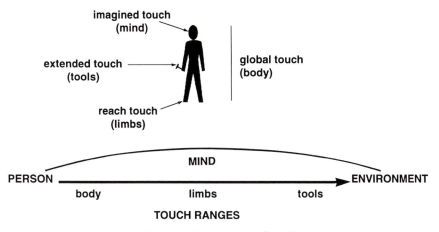

Figure 4.3 Dimensions of touch

53

spatial orientation. The microscope and telescope substitute a visual impression for a touchable surface. Furthermore, materials covering the body may mediate the relationship to the tangible world (see Montagu 1971). The everyday practice of wearing shoes and other protective clothing can limit touch information or modify it. Protective clothes may allow us to touch things which would ordinarily damage the skin, yet feeling through gloves is never the same as direct touch exploration of a surface.

Finally, it is possible to identify an imagined touch, that is a haptic experience rooted in the memory and expectation. This is demonstrated both in our use of touch metaphors to describe other sensuous experiences and in the creative recall of haptic experiences, as when reading a description in a novel or when remembering a treasured experience. This rich touch imagination permits us to experience an intimacy with people and places which may be a great distance from our present location, in time and/or space, or which we have never actually experienced, such as the evocation of tactile experiences in dreams or when reading. 'Keeping in touch' is most often, it seems, referring to letter writing or, nowadays, the ubiquitous phone call, rather than a literal connectedness with other people and places.

Haptic experience can therefore generate a range of distinctive geographies, giving both much detail about our own bodies, objects touched and the wider environment in which they are situated. This geography is both passively and actively encountered, involving the tactile receptivity of the skin, the mobility of the body and the body as measure. Focusing on the dimensions of touch in individual experience also reminds us that this geography is always, ultimately, in reference to a human body, our body, and each space and place discerned, or mapped, haptically is in this sense our space and because of the reciprocal nature of touch we come to belong to that space. In this sense, the sense of place is grounded in the participatory quality of haptic geography, both degrees of insideness and outsideness in the relationship to places (see Relph 1976).

Table 4.1 Touch words

touch (material, metaphor)	grab (take hold of)	stroke
contact	hold	rub
feel (specific, general)	weigh	caress
reach	pick up	snuggle
explore	press	squeeze
etc.		

HAPTIC GEOGRAPHY, CULTURE
AND SPACE

In different social and historical contexts touch plays different roles in interpersonal and environmental relationships. These alternative relationships to touch are sometimes, though not always, reflected in differences in the organisation of the built environment. However, one must be careful in over-generalising about apparent cultural differences in touch and in speculating about the consequences for the design and organisation of space. There are many variations within a cultural group. Furthermore, the link between the style of haptic experience and the organisation of space is not a simple one.

1   It may be a *chicken and egg situation* – which came first: the tactile style or the spatial organisation?
2   The built environment is a *historical artefact* reflecting past behaviours as much as, if not more than, present ones (which might have changed markedly), but we 'make do' with existing structures.
3   There is *no exclusive one-to-one relationship* between touch (or haptic sense) and environmental design. Design arises out of a multisensuality of experience conditioned by economic and social factors, one has both aesthetic and functional influences.

Within Europe there are a variety of styles of touch behaviour. Frequently, anecdotal evidence is quoted to suggest that southern Europeans have a greater propensity to tactile activity in interpersonal relationships and do much hugging and kissing in public transactions as well as in personal relationships. In contrast, it is suggested that northern Europeans are less tactile in their interpersonal relationships, and especially the British, favouring limited body contact in public transactions (largely restricted to formal hand-shakes) and restricting general touching and hugging to private and more intimate familial or romantic contexts. This haptic difference may also explain differences in attitudes to relative crowding of people in spaces and built environment preferences (Evans and Lepore 1992; Aiello *et al.* 1981; D'Atri 1975; Stokols 1976; Evans and Howard 1973).

Sometimes a distinction is made between contact and non-contact people and cultures (Hall 1969; Tuan 1993). This distinction perhaps should be described as a continuum rather than an absolute contact/non-contact alternative and certainly in different contexts – such as public versus private spaces – tactility can vary within a culture. These varieties of touch behaviour can be observed in both interpersonal relationships and in the organisation of space in the built environment. To fully understand such differences in haptic geographies one needs to recognise the interlinkages between the senses and the broader cultural values operating in a given context. Such an anthropology of sensuous geographies is not attempted here, but in the space available some key features of these cultural differences will be sketched for Japanese, Arab and North American contexts.

Japanese space

The Japanese offer a complex and poorly understood illustration of alternative attitudes to the environment, concepts of space and place, and manifestations of sensuous experience (Omata 1992; Arakawa 1988; Canter and Lee 1974). This is demonstrated in their poetry (especially the famous haiku), the traditional gardens and house forms, and contemporary attitudes to urban space (Ogawa 1980; Johnson 1989, 1990; Omata 1992). Some of the explanation of the Japanese definition of the senses, sensuous experience and space can be understood by considering traditional attitudes to nature and environment, but one must also consider the impact of contemporary cultural values (Stanley 1988; Hall 1969).

In the traditional Japanese house, 'the collective human relation was reflected in the spatial organisation of the house, and the rooms for guests or rituals were more important than those for family members' (Omata 1992: 259). After the Second World War, Japan changed in many ways and this is particularly evident in life style and the form of houses and their internal spatial organisation, especially in urban areas. Private rooms became important and these were given Western-style doors and walls (Hagino *et al.* 1987). The Japanese-style sliding door (*fusuma* or *shoji*), which was ideal for the flexible use of space, was abandoned in the newly designed apartment blocks. As part of a general Westernisation of lifestyles, especially in larger urban areas, even the traditional straw floor mat (or *tatami*) was replaced by Western-style floor coverings (Kikusawa 1979). All these changes fundamentally altered the nature of Japanese living space, its appearance, its auditory resonance, its olfactory dimension and its tactile character. The new apartments and houses – more Western than Japanese – were far less flexible spaces than the traditional Japanese-style dwellings, and more compartmentalised – with rigidly demarcated rooms and, increasingly, the family members organising themselves between such rooms in more Western-styles of personal privacy and individuality. These new structures contradict the traditions of Japanese culture and its fondness for close physical bonds between people, especially within the family unit and community.

In the haptic geography of the Japanese living space we can observe two key changes in the transition from traditional dwellings to modern Western-style apartments and houses, for instance:

1   In the traditional Japanese dwelling, reception space (for entertaining guests) was separate from the family living space and occupied the best space. (This is not dissimilar to the traditional English use of the day-time spaces of the house: the back-room or kitchen/eating area often

56

formed a 'living room' for the family, whilst the front room or 'best room' was reserved for Sundays and for entertaining guests. Since the Second World War, this demarcation of space has tended to disappear.) In modern Japanese houses, these two spaces are often somewhat overlapping, that is, these needs are served by a common space or no longer are the two spaces preserved exclusively for the separate functions (Omata 1992).

2   In traditional Japanese dwellings there was no private room for each family member, and really little need for such rooms since it was common for families to sleep together, either quite literally or at least within the same 'bedroom area'. In the modern Japanese house, there is a much greater tendency for family members to sleep in separate rooms, especially teenagers and single adult family members. Omata (1992) reports on research which indicates marked separation of the space of children and parents.

The haptic geography of the modern Japanese house is thus quite different from the traditional dwelling (see Ogawa 1980). Personal space behaviour (or tactile distance) has become more like that found in the West. This is evident both in the separation of sleeping spaces and in the separate demarcation of day-time space. A tactile distance in social relationships and greater privacy, both within the family and from the community, is increasingly expected by urban Japanese. Interestingly, Omata (1992) observes how Japanese house-holders increasingly lock their doors and show concern about security.

Arab space

It has long been recognised that, traditionally, Arabs organise personal and social space quite differently to Europeans and North Americans. This is seen in the traditional Arab city and the courtyard houses (Norberg-Schultz 1980; Serageldin and El Sadek 1982). The haptic geography of Arab cultures is changing, but nevertheless, many of the traditional patterns of behaviour are preserved, especially in interpersonal relationships and attitudes to personal space (Hall 1969). Researchers have become increasingly aware of the impact of modernisation (or Westernisation) on the traditional Arab or Muslim city and the conflicts that arise between Western styles of building and planning ideology and the traditional approaches (Bianca 1982; Cantacu-zino 1982).

Hall (1969) observes how for the Arab body-contact is a vital part of interpersonal relationships, especially in public transactions. Both much touching and proximity of bodies is seen as vital in establishing a relationship of trust and in securing a business transaction or, more generally, in sharing a friendship. Arab interpersonal relationships clearly illustrate the important reciprocity of touch and the directness of tactile interaction with its association of truth and commitment. For the Arab, the ability to smell the other person and to be smelled by them is important in the exchange and to hold a distance – arm's length or beyond – where smells cannot be exchanged is to express a lack of trust, even of hiding something. Thus we cannot fully understand this haptic geography of interpersonal relationships without also taking into account the olfactory dimension.

This basic difference from Western preferences in interpersonal distances and tactile behaviour, is also reflected in the spatial organisation of traditional Arab urban space. A fundamental conflict has emerged in Arab urban space between the values of Islam, which traditionally demand certain patterns of social behaviour and organisations of space for specific activities and demarcations of space functionally (e.g. between trades) and socially (e.g. between the sexes) and Western science and technology, which has also introduced Western values (Bianca 1982).

The Westernisation of Arab urban space transforms haptic experience at the social and economic level. For instance, Western planning methods emphasise the division of space along functional lines and assume interaction between economic activities can take place over greater distances. Hence, greater use of vehicles is needed and traffic flows across the urban space increase and telephones and fax machines, it is assumed, will substitute auditory for tactile-auditory encounters between the agents of economic activity. Arabs, however, feel a loss of social communication in this urban design – interpersonal trust is not so easily generated over the telephone as in person-to-person contact. 'The traditional Islamic city stresses the idea of close interrelation of the various aspects of urban life' (Bianca 1982: 40). This is not to deny that the traditional city did not have spatial divisions on ethnic and religious lines, but within a social group close proximity was valued.

Also, in the traditional Arab city there is a strong distinction between private and public space which reflects a concern for privacy in private residential quarters (Hall 1969; Bianca 1982). However, though the courtyard houses with their 'back' on the street seem to isolate themselves from the city around, this traditional residential space is very much nestled within the community and structure of the city, and contrasts with Western style houses in wide streets with gardens and drives, windows exposing occupants to the street, or tower blocks rising to the sky – each form being experienced by the Arab as far more isolating and detached from the community of the city. In the Arab context, privacy does not mean distance from neighbours, merely a division of space into private and public areas.

In public life, Arabs seem to favour a far more immediate haptic geography than Europeans or North Americans. Relatively small and crowded spaces are characteristic of the permanent markets of inner city, where people crowd together to secure genuine transactions in the congested souks and bazaars. The Arab is highly tolerant of high densities of people and much bodily contact in the crowded souks (markets) and alleys of traditional cities. Yet, in private spaces, such as the home, once greetings have been intimately exchanged with hugs and kisses, friends may sit in quite large rooms often at opposite ends of it to talk. This spatial organisation reflects trust and friendship which has already been established (Hall 1969).

The traditional Arab city has an urban fabric not of wide boulevards and streets, but of narrow alleys and small courtyards. It is a cellular rather than linear structure (Cantacuzino 1982). This reflects and reinforces a geography which is more tuned to haptic experience than Western urban spaces. In cities such as Kairouan (Tunisia) one readily sees the contrast between the souk quarter and the French colonial quarter of the city.

American space

Americans have been noted for maintaining certain clear interpersonal distances in different kinds of relationship (see Figure 4.4). Hall (1969) suggests that Americans largely restrict close contact to intimate relationships, especially those between friends and relations. In public spaces, he suggests, there is a preference for 'personal space' of a minimum distance around about 1.5 to 4 feet within which body contact is avoided. The crucial distance was studied in depth by Sommer (1969) who defined personal space as an emotionally charged space bubble around each individual which is regarded as private and personal territory. It is effectively an extension of the self's presence in space and violation of this space, or territory, by another is felt to be like the violation of the body itself (Sommer 1969; Evans and Howard 1973).

Americans appear to hold very strongly to this personal space when in public areas, even moving into open space when others enter the area (Hall 1969). This is also seen in the way in which people tend to sit on alternative seats or at a distance from one another on public seating. In a non-contact culture, people feel crowded if they need to stand closer to the individual

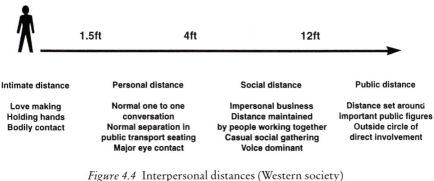

Intimate distance	Personal distance	Social distance	Public distance
Love making Holding hands Bodily contact	Normal one to one conversation Normal separation in public transport seating Major eye contact	Impersonal business Distance maintained by people working together Casual social gathering Voice dominant	Distance set around important public figures Outside circle of direct involvement

Figure 4.4 Interpersonal distances (Western society)
Source: based on Walmsley (1988: 91)

59

with whom they have no intimate relationship (Baum and Koman 1976). Hence public spaces are large and open, such as large hotel foyers, roomy airport lounges, broad corridors in buildings, boulevards and large open areas in public parks are all common features in the public spaces of American cities.

In private spaces, such as the home, interpersonal contact is generally more frequent and spaces are much smaller without any sense of crowding. However, the lounge – where visitors are entertained – is often the largest room in the house. In the home, smaller spaces may indeed be described as cosy and intimate, favoured for their friendliness and privacy. Here is also an example of the complex interlinkage of the senses. Personal space appears also to be reinforced by specific attitudes to body smells. An American prefers not to smell the body odour of another person nor to feel that person can smell their body odour. Personal space is distant enough to allow no olfactory communication except specially chosen scents and perfumes which project a particular image and further mask natural body smells.

Therefore, in briefly observing Japanese, Arab and American behaviour in space we begin to appreciate important differences in haptic geography, both at the level of interpersonal relationships and at the level of the organisation of urban space. Furthermore, closer examination reveals that this haptic geography – or a geography of styles of spatial distance and contact – is not merely a tactile issue. In fact, to fully describe each of these cultural differences in the organisation of space, we have to consider more carefully the multisensual nature of geographical experience. Arab attitudes to personal space, for instance, are closely linked to a more positive attitude to olfactory experience. The American behaviour seems, in part, to reflect a more negative attitude to olfactory experience and a greater reliance on visual cues in the environment over tactile ones in interpersonal relationships and in organisation of social and economic activities.

5

OLFACTORY
GEOGRAPHIES

INTRODUCTION:
A SMELL–TASTE SYSTEM

Henri Lefebvre writes: 'where an intimacy occurs between "subject" and "object", it must surely be the world of smell and the places where they reside' (1991: 197). Olfactory geographies are like haptic geographies, both are quite intimate and immediate yet ordinarily much neglected as our attention is drawn to the geographical knowledge generated by the eyes and ears. Yet, without the haptic and taste–smell system, much of our ability to locate ourselves in space, distinguish friend from foe, recognise food and identify ourselves with a home-space would be lost. Olfactory experience – the geography of the nose – provides an important dimension of sensuous experience.

Lawrence Durrell wrote: 'losing one's sense of smell ... why is there no word for it in English? We have deaf and blind, but nothing to describe the lost sense of smell ...' (1992: 110). Of course, there is today a technical term used to describe the loss, partial or wholly, of the sense of smell. This is anosmia. 'Little attention has been paid to [this] ... it is possible for humans to be unaware of an olfactory deficit' (Engen 1982) – a telling observation. However, Durrell's observations remind us of a tendency to neglect the sense of smell. Engen, as a scientist specialising in olfactory perception, also notes how limited research has been until very recently and especially in comparison to research into visual and auditory perception. Olfactory disorders or losses appear to be more widespread than we realise, from anecdotal stories about the person who cannot smell the escaping gas or the beauty of flowers, to scientific measurements of olfactory sensitivity to specific odours (Engen 1982; Gilbert and Wysocki 1987). Whilst a physiology of smell, or olfaction, can be described, it is also clear that each culture (and period) defines olfactory experience differently and its role in geographical experience (Hall 1969).

Few geographers appear to have considered the olfactory dimension of geographical experience (Tuan 1974, 1993; Porteous 1985, 1990). Never-theless, a number of non-geographical studies provide valuable insights into the history of olfactory research (e.g. Boring 1942; Cain 1978), olfactory physiology and odour perception (Engen 1982), smell preferences (Mont-crieff 1966; Engen 1972), and the artificial nose (Gardner and Bartlett 1992). The smell survey conducted by the *National Geographic* also provides valuable insights into the scope of olfactory research, the variety of odour experiences, and individual and cultural differences (Gibbons 1986; Gilbert and Wysocki 1987). An olfactory geography would, however, be primarily interested in the role of smell in geographical experience, that is in the organisation of space and spatial relationships, locatedness and orientation in space, and the characterisation of and relationship to place.

Porteous (1985) has popularised the term 'smellscapes' to describe the geographical dimension. Here, the term 'olfactory geography' is chosen in preference. Olfactory is a precise technical term used to refer to the sense of smell. It is inclusive of both the activity – the verb, 'to smell' – and object – the nouns, 'smell', 'odour' or 'aroma'. Smell is supposed to be a neutral term, but in practice it often has a strong negative connotation. 'It smells' generally means 'It has a bad smell.' Odour is also more often used in its negative sense. In contrast, aroma tends to be most used as a positive olfactory experience. 'To smell' can refer to both actions such as sniffing a flower and the situation of an individual or object giving off a smell. The term olfaction transcends this ambiguity and concentrates attention on the relationship which subsists when smells, odours or aroma permeate space (whether between people or people and things in an environment).

Gibson refers to a smell–taste perceptual system (1968). Whilst in the present chapter emphasis is placed on the sense of smell and the sense of taste is not specifically considered, it is important to remember that the two senses are closely linked and, certainly in the context of the enjoyment of food, are strongly linked so making meaningful distinction difficult, if not impossible. Both taste and smell appear to be chemical senses. Taste seems to be associated primarily with the taste buds on the tongue, and smell with the hairs in the nose. Taste is associated with particles suspended in liquid, smell with particles suspended in air. Gibson (1968) suggests that a useful way to distinguish them is by their function: smell accompanies breathing and taste is associated with eating. In other words, taste monitors the intake of food and smell monitors the intake of air quality. Yet taste always seems to also be implicative of smell, and olfaction plays a key role in finding and differentiating foods. Just as taste always seems to have a smell behind it, so to speak, so smells seem, on closer reflection, to more often than not have a taste dimension. The strongest odours seem to 'attack' the tongue as much as the nose. Here, olfactory experience is concentrated upon since it appears more immediately geographical than taste. For taste to operate things have to be placed on the tongue or in the mouth, but smell gives us access to a world around us, a wider array of stimulation some of which we choose to sniff but much of which bombards us, that is, we encounter without choice. A more complete geographical analysis would also include taste, but this should not be a geography of 'taste and manners' but rather a more intimate geography of encounter and spatial orientation through the use of taste – itself a quite rare phenomenon. In any case, by identifying a taste–smell system, it implies that smell is often implicative of taste perception. Taste could even be described as a kind of touch. Here, olfactory perception is explored as the more 'visible' geography of this system.

Porteous (1985, 1990) uses the term 'smellscape', which is analogous to the term 'landscape'. In particular, he considers the rich detailed evocations of place and attachments to specific places found in novels, biographies, poems and diaries. Smellscape has much the same limitations as the term soundscape (see Chapter 7), since the analogy to the term 'landscape' brings with it the connotations of artistic creation and aesthetic contemplation. There is much

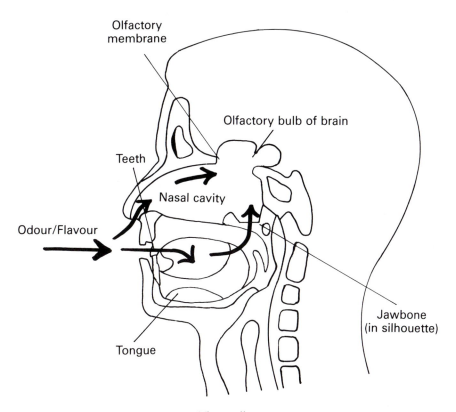

Figure 5.1 The smell–taste system
Source: adapted from Gibson (1968: 139)

more to olfactory geography than this relatively passive view might suggest. Nevertheless, Porteous's concept provides an important starting point for critical research into olfactory geographies and it is important to note the scope of interest in smellscapes he identifies. Smellscape is intended to suggest that 'smells may be spatially ordered and place-related' (Porteous 1985: 369). In itself, this is quite plausible but needs fuller explanation.

The term '-scape' implies there is a relationship between person and environment (Houston 1982). It is this relationship both intimate yet illusive that Porteous focuses on in the 'pictures' of olfactory experience, rather than the olfactory relationship with the environment and its wider function. Smellscape as an equivalent of the visually grounded term 'landscape' (see Tuan 1974; Cosgrove 1984) fails to genuinely reflect the everyday nose-experience of a world around us. Smells do not offer scenes or views, objects arranged and set at a distance from the observer. Rather, smells are present or not present, in varying degrees of intensity and subject to the movement of air (or the locomotion of our bodies through space). Smells infiltrate or linger, appear or fade, rather than take place or situate themselves as a composition.

One can recognise a smell far more easily than one can name it or describe it to others (Lawless and Engen 1977). The bottle of perfume only has a superficial resemblance to the painting offered by the artist. Some novelists have nevertheless provided rich evocations of places and people through detailed olfactory descriptions (e.g. Herriot 1977; Greene 1971a, 1971b; Huxley 1978). Smells offer a complex and dynamic geography, both of current olfactory experience and odour memories and associations. The present chapter explores the nature of this olfactory geography.

OLFACTION, SPACE AND SMELLSCAPES

The olfactory sense has several important characteristics.

1   It enables us to identify distinctive odours and associate them with particular sources (things, organisms) and/or situations. Olfaction offers a sensuous geography, sometimes described as 'smellscapes'.
2   It is inclusive of a physical sensation (chemical) and mental activity (thinking), and odour memory is especially important, being often long and accurate. Olfaction plays an important part in remembering in general and the association of current and past place experiences.
3   It is an adaptive sensitivity, which is excited by novelty but dulled by familiarity, or habituation. Perhaps in this sense, olfaction has an important warning function by drawing our attention to change in the environment.
4   It is strongly associated with the emotions and the encounter with specific smells and smell intensities excite particular emotional responses

– though such correlations are not always simple nor can be generalised for all individuals or cultures.

Each of these characteristics is closely interconnected and this suggests that olfactory geographies are quite complex, including reference to relationships between sources and effects, across time and space.

Study of olfaction and smellscapes is, however, constrained to some extent by the limitations of our language (Porteous 1985). Olfactory researchers have also noted this language problem (e.g. Lawless and Engen 1977), but attempts have been made to develop a technical vocabulary (Engen 1982). However, in everyday experience inevitably we often refer to specific smells not by unique words but various common (and sometimes more creative) situation associations and metaphors. Thus we refer to the 'smell of freshly cooked bacon' or 'the smell of ground coffee', more often than a 'bacony' or 'ground coffeeish' smell. We describe an odour as 'like fresh mown grass', 'like the smell of the sea', or 'like the morning after the night before'. The fact that we don't always have the vocabulary to give our olfactory descriptions precision or that we find it difficult to name a smell does not necessarily mean that we are not quite acute at distinguishing different smells. This difficulty of naming smells is sometimes known as the 'tip of the nose problem' (Engen 1982; Lawless and Engen 1977). Although memory of smells is often remarkably acute and has great longevity, we are more able to recognise smells than to recall them, to match smells to a list of suggested names than to actually think of a name (Lawless and Engen 1977). Of course other animals, such as our pet dogs, seem to have far more acute senses of smell (Boring 1942) yet they have no vocabulary of smells. We clearly do recognise key smells when they are at appropriate concentrations, or densities, and readily compare smells, identifying differences and similarities and associating different smells with different features in our environment.

The argument about language is, however, more complicated than vocabulary. It is also a grammatical or structural problem. The subject–object dichotomy of everyday language forces our description of olfactory experience into an inappropriate framework. Smells are not neatly defined objects in the sense of visual objects but experiences of intensities, more like those of pain and joy. It is for this reason that Lefebvre (1991: 197) can argue that smells cannot be inventoried, for no inventory of them can have either beginning or end – except life and death. Many attempts have been made in the past to classify odours, but despite some recurring labels no accepted inventory has been established. Porteous (1985) observes that the classifications range from four to as many as forty-four, with mainly around seven basic types of odour identified. It is interesting to observe how similar Linnaeus's classification of 1756 is to the one offered by Amoore over 200 years later (Table 5.1).

The inadequacy of such descriptions is immediately obvious by the

Table 5.1 Odour classifications

	Linnaeus 1756	**Amoore 1970**
1	aromatic	ethereal
2	fragrant	floral
3	ambrosial (musky)	musty
4	alliaceous (garlicky)	pepperminty
5	hircine (goaty)	camphoraceous
6	foul	pungent
7	nauseating	putrid

Source: Engen (1982: 8), Porteous (1985), Gibbons (1986)

exclusion of many everyday olfactory experiences. It is interesting to note how several of these types of odour have strong taste associations. This reinforces the argument that taste and smell should be treated as part of the same perceptual system. More specifically, what makes these seven (Linnaeus or Amoore) significant? They seem to mix up a number of aspects of olfactory experience. For instance, nauseating suggests an effect of odour, whilst ethereal and fragrant each suggest a characteristic of many odours, and yet other categories, such as peppermity or hircine more closely refer to a specific smell. Moreover, not only are the smell labels of mixed type (effect, character, specific) but also arbitrary and do not explain or assist an explanation of olfactory experience.

These smell classifications do not compare with the powerful typologies of light and sound. The chromatic and modal divisions of auditory intensities, or pitches, are quite arbitrary and different cultures have favoured different divisions of the scale in their music-making. For instance, the harmonic scales of Europe contrast with the traditional modal scales of the Indian subcontinent. In this sense, sound is not an object like visual objects but a continuum of varying pitch densities which by a specific cultural tradition is subdivided into a scale of notes. Odours seem also to be a continuum of intensities (quantities?) but seem to vary in many qualitative ways too. Many cultures have recognised key odours or odour types, but nothing analogous to an 'odour scale' has been established.

Nominal classifications of 'primary' odours are analogous to the naming of the primary colours of blue, green and red. However, the real value of the primary colours is not so much the specific colours or labels (which differ from culture to culture) but that they can be defined in terms of the physical correlate of wavelengths, such as 460, 530, 640 μm respectively. By mixing these wavelengths one can produce any other colour of the visual spectrum (pitch frequencies operate in a similar way for sound). Fitting particular smells to primary categories has proved extremely difficult despite careful chemical research. Closer study of these 'primary' smells themselves fails to reveal physiochemical correlates analogous to wavelengths. The odour problem appears to be more complex than the colour problem. It is no wonder, therefore, that Boring (1942) estimated in the 1940s that olfactory research was still largely at the stage equivalent to visual or auditory research in 1750. The problem of description remains dominant in olfactory research and scientific understanding is still limited and fragmentary.

Smell in geographical experience is complex, including both immediate encounter with the environment and a kind of virtual encounter with places in the imagination when odour memories are excited by current place experiences. Olfaction seems to offer a time–space geography, both at the level of current durations of odours in space and in the lingering of odours in memories. This geography, however, does not seem to be an experience of distant space nor an experience of detachment from what is perceived, as in the case of visual and auditory experience. Rather, olfaction is concerned more with encounter with a near or immediate environment and an involvement in that environment, as with haptic experience. It is indicative of both the close association of smell and taste, and the intimate geography of olfaction, that we feel the taste of a strong odour as much as we smell it, and strong smells seem almost tangible to us, as if we could 'touch' them. The chemical basis of olfaction makes it a far more direct sensuous geography than sight, and perhaps also hearing.

Smell has therefore been described as the intimate sense (Gibbons 1986; Tuan 1974; Gold 1980). This is intimate in two senses: structurally – it generates an immediate or local geography – and emotionally – it establishes a strong bond between person and environment. As with haptic experience, this bonding is established by the direct contact between the body, or the smell–taste perceptual system, with the environment or features within it. This is a direct chemical contact. The more abstract senses – sight and hearing – do not seem to have quite such direct contact. We might be able to close our eyes but we cannot close our noses.

Further, familiarity with a particular odour – known as habituation (Engen 1982) – may dull our sensitivity to the present encounter with it, but in another time and place the fresh encounter excites a detailed recall of the odour and its associated experiences specific to our own biography.

One can begin to decipher olfactory geography by remembering the everyday use of the verb 'to smell': it can refer to 'smelling a rose', that is, perception, or to 'having a smell', that is, having a presence. This is illustrated in the olfactory matrix (Figure 5.2). One can identify a kind of touching, or rubbing against, which is passive and not active. However, it is more difficult to think of an olfactory experience which is merely juxtaposition. All olfactory experience involves some kind of interaction, a chemical encounter between odour particles in the air and sensory membranes in the nose. Olfaction – in this sense like the rest of haptic experience and auditory experience – requires also some kind of movement by definition. The distinction between active and passive here, therefore, is between unintentioned or stationary sentient and

sensation

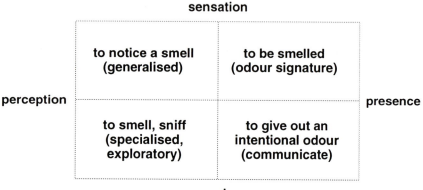

| perception | to notice a smell (generalised) | to be smelled (odour signature) | presence |

to smell, sniff (specialised, exploratory) | to give out an intentional odour (communicate)

meaning

Figure 5.2 The olfactory matrix

active or exploratory sentient. Expressed alternatively, olfaction is generally interactive, a kind of communication with the environment – sniffing and/or giving out a smell – and fundamentally this occurs across space, hence it is geographical.

The basic geography of smell is therefore experienced as the perception of an odour in or across a given space, perhaps with varying intensities, which will linger for a while and then fade, and a differentiation of one smell from another and the association of odours with particular things, organisms, situations and emotions which all contribute to a sense of space and the character to places. We may not be able to name the smell – the 'tip of the nose problem' (Lawless and Engen 1977) – but this does not prevent us from recognising it and reacting to it. Habituation – that is, familiarity – generally reduces our perceptivity to olfactory geographies but, by contrast, we are more receptive to the appearance of smells and especially to newly encountered smells.

This olfactory geography may not generate a 'map' in the traditional visual sense of a synthetic overall view of various odours, their source, location and the general organisation of olfactory space. Yet, Porteous has reflected that 'unlike touch, however, smell does not seem of great value in structuring space'. Its strength seems to lie in place evocation and the remembering of places (Porteous 1985). In fact, even though a given olfactory encounter may or may not extend to discerning a source for the odour or its location, it can still be important in giving us qualitative information about the character of the environment and people and things in that environment. Porteous (1990) argues that olfaction is more important in place evocation than in discerning spatial structure, whilst Bekesy (1964) suggests that the sense of smell, like hearing, is in fact keen in its ability to locate stimuli. This apparent contradiction is clarified when we realise that we often have to sniff out the precise source of an odour, or smell, as a dog sniffs along a scent, in order to locate the smell.

Therefore, two styles of olfaction can be identified (as in Figure 5.2). First, *a generalised olfaction* – a kind of passive encounter with odours in the environment giving an imprecise sense of their location but much detail about their qualitative character. If these odours are innocuous or familiar, we soon forget or ignore them. Second, *a specialised olfaction* – this is characterised by exploratory behaviour which is excited by certain odours, intensities, associations or memories. This exploratory olfaction tends to focus-in on specific smells, rather than attempting to compose an overall smellscape.

These two styles of olfaction suggest that the spatial structure discerned by olfaction is not so continuous, integrated and clear as the visual, auditory and tactile space. In this sense, we do not experience smellscapes (analogous to landscapes and scenes) but do discern discrete 'episodes' or 'events' of olfactory encounter which generate a more discontinuous, loosely associated geographical experience of current, remembered and potential olfactory encounters. However, both styles of olfaction do give us much detail about the character of our environment, and substances within it. Objects or agents may emit particular odours of certain intensities which spread across space (or through the air), but these odour 'clouds' move much as we might move in exploring the environment, dilute or dissipate once the original source has ceased to emit, and can become contaminated by other smells including our own body odours and perfumes. To discern a 'map' or smellscape of such an olfactory geography is probably both impossible and unnecessary. Of course, in our imagination and in the place descriptions of novelists, rich smellscapes can be synthesized but this is not how we actually experience smells in everyday experience, where they appear and disappear as discrete 'events'. This is not dissimilar to the musical score which gives a synthetic view of the whole piece of music, though we experience its performance as a series of unfolding sounds.

Although our sense of smell is generally regarded as less acute than that of many other organisms, its main geographical role seems to involve differentiating odours, associating them with particular needs (such as recognising foods) and importantly as a kind of warning mechanism.

Habituation means that our sensitivity to smell rapidly declines with exposure and the first impression generally proves to be the most important in establishing a response behaviour and in developing odour memories (Engen 1982). Habituation is classically illustrated when one visits another person's home for the first time or in the visitor's impressions of foreign land. The evocative smells which disturb or excite the visitor, may not be ordinarily noticed or valued by the resident or local. Only perhaps when travelling away from home or some years later when encountering once familiar smells again does the olfactory experience become again richly experienced. Engen (1982) suggests that this insider/outsider or familiarity/novelty distinction is the basis of the assumption that olfaction once played a key warning role essential to biological survival. It also reinforces the earlier suggestion that olfactory geographies are not consciously continuous, unlike visual experience, but discontinuous, like a series of events (or discrete encounters) such as those consciously identified with actively touching and in the recognition of distinct sounds in the environment.

The longevity and detail of odour memory has long been recognised by writers of biographies and novels and by scientific researchers. Lawless and Engen (1977), for instance, have investigated the characteristics of odour memory and the ability to name odours. They found that odour memory is long term rather than short but is acquired more slowly than verbal and visual memories. However, these memories are very detailed, accurate, long-lived and seem unpolluted by subsequent experience. The vivid childhood memories of the smells of particular places and their emotional associations so often reported (Porteous 1985; Hart 1979) and the not infrequent experience of *déjà-vu* which places can suddenly evoke in our consciousness, each attest to the importance of this olfactory memory. Tuan (1974, 1993) reminds us how humans – and especially contemporary North Americans and Europeans – are predominantly visually orientated and sight appears to be our dominant sense, but olfaction (and the other senses) can still be vitally important and greater understanding of this may undermine such assumptions.

Each of the senses has emotional correlations, often ranging through positive and negative responses of varied character depending on the quality and intensity of the sensuous experience and the context of that experience. Olfaction is particularly evocative of emotional responses. 'Smells are surer than sounds or sights to make the heart-strings crack' (Gibbons 1986: 324). It is difficult to test this kind of assertion, but it is clear that olfactory memory and emotions are closely associated, and emotional responses are an integral part of the warning/differentiation function of olfactory perception. Lefebvre (1991) suggests that olfaction is in someways pre-reflective and pre-rational, and offers a more direct and unpremeditated encounter with the environment and its features. Certainly, it can be observed that young children have a particularly strong attraction to smells (and tastes) and this forms an important part of their early explorations of the mother and subsequently the environment. Many of these child odour preferences the adult finds extremely distasteful, especially in Western cultures where we tend to value more subtle odours over pungent or strong ones. Two elements seem to explain this difference between the olfactory geography of the young child – rich and exploratory – and the adult – poorer and more passive. First, there is some evidence to suggest that almost all the senses are experienced more acutely when young, either because our sense organs are more effective or because many experiences are new and fresh, so exciting and interesting to explore. The deterioration of sensitivity to olfaction is, however, much less in reality than that commonly experienced for hearing and sight (Engen 1982). Second, and perhaps more importantly, there is a socialisation of perception. We are conditioned by education into the cultural practices and attitudes to odours shared by the culture to which we belong.

Therefore, olfaction gives a distinctive sensuous experience of space and duration; past, present and potential spaces; and this is both physiologically grounded and culturally defined. Olfactory geographies are not merely 'smell maps' or even 'smellscapes', but complex emotional encounters with discrete olfactory events, odours passing through time as well as space. Immediate olfactory encounters are perhaps quite a discontinuous geography, but the rich evocations of remembered smellscapes suggest that underlying this is a more continuous and even synthetic olfactory geography. Habituation fragments immediate olfactory experience, memory connects it together and situates it within a myriad of associations. Olfactory experience is, however, not mere juxtaposition but always a relationship, chemical or mechanical, between that which (gives off) smells and the individual who smells (or sniffs).

EXPERIENCES AND METAPHORS

Although, in everyday experience olfaction is often ignored or given secondary place to other sensuous experiences, we nevertheless do value smells and give olfaction specific roles in geographical understanding, current and remembered. As we have already noted, olfactory memory is an important quality of olfactory geographies. This is especially evident in the memories of childhood (Hart 1979; Porteous 1990; Tuan 1977, 1993) and in the place-evocations of novelists and poets and in their descriptions of the character of people and things and our relationship to them (Chesterton 1958; Ruark 1964; Bates 1969; Greene 1971a, 1971b; Huxley 1978; Herriott 1977; Malancioiu 1985; Rice 1991). Porteous (1985: 369) notes how the novelist Graham Greene has admitted that 'smell to me is far more evocative than sound or sight'.

In describing olfactory experience, and in remembering such experiences, we often make associations and employ metaphors. In English, there is relatively limited specific olfactory vocabulary, especially for olfactory geographies, but we nevertheless find ways to share our experiences with others and to remember them. Poets frequently cannot offer a precise smell word for a given odour but can still evoke situation, experience or place by the reference to an association: Malancioiu evokes the odour of a hot racing horse: 'that flesh under my saddle smells of blood' (1985: 97). The association to blood evokes both a well-known smell but also one symbolic of life and death.

Ruark captures the associative power of olfactory experience, both of places, people and our attitudes to them, time and again in his *Uhuru* (1964):

> the smell of white man, the white man's food and drink and clothing, the greasy stink of the white man's petrol fumes and belching diesel exhausts ...
>
> (149)

> the howling, reeking bazaars, where every smell known to the East was mingled in one magnificent ripe stink of rotting fruit and dust and dung and curry powder and wet plaster and no plumbing and ancient filthy habits.
>
> (97)

And yet, despite the richness of such olfactory evocations, they can seem more archetypical than specific, more of a kind of experience, association or metaphor, than fixed to a particular experience or place encounter. This is evident in the accounts of childhood smells in John Raynor's *London*, which Porteous (1985) considers in some detail. Raynor's evocations often suggest archetypes rather than just personal biographical details. For instance:

> Suddenly on the wind was borne the smell of the tannery. I stood, transfixed, dropping my flowers, turning green and white, gripped by ... a horror so primitive that it would only have been a racial memory; a horror quite outside the bounds of thought or control; something that struck deep inside my body ...
>
> (quoted by Porteous 1985: 372).

The olfactory geography presented by novelists and poets is not just dependent on their art as describers of experience, it is also dependent on the reader's own experience and ability to imagine the olfactory experience. Olfactory experience is very personal. The evocation of a childhood smell is first and foremost 'my childhood places'. Unlike colours and sounds, we do not seem to have a precise set of learnt odours in our heads, a vocabulary of our experience in this sense. There is no real olfactory equivalent of the cinema or symphony concert, yet the international perfumery industry is beginning to establish certain odours as widely recognised fragrances with specific associations through mass advertising. The nearest the writer can get to sharing an olfactory geography is to remind us of our own experience by some common association or emotional response, perhaps linking it to childhood or widely known stereotypes of odour associations – school books and libraries with a musky smell; the sweaty smell of the changing rooms for sports; the crisp, fresh smell of the newly washed and starched linen of the dormitory; and of course the odours of first love. Malancioiu evokes the rich olfactory memory of a place in just two lines, hoping the reader has experienced this rural smell: 'fresh air with a smell of dug-out vine and with whitewashed trees at the gate' (1985: 99).

Yet, smell memory can be quite illusive, as for instance the experience reported by Edward Thomas of the scent of 'Old Man or Lady's Love', a hoar-green feathery herb:

> As for myself,
> Where first I met the bitter scent is lost.
> I, too, often shrivel the grey shreds,
> Sniff them and think and sniff again and try
> Once more to think what it is I am remembering,
> Always in vain. I cannot like the scent,
> Yet I would rather give up others more sweet,
> With no meaning, than this bitter one.
> I have mislaid the key. I sniff the spray
> And think of nothing; I see and hear nothing;
> Yet, seem too, to be listening, lying in wait
> For what I should, yet never can, remember:
> No garden appears, no path, no hoar-green bush
> Of Lady's-love, or Old Man, no child beside,
> Neither father nor mother, nor any playmate;
> Only an avenue, dark, nameless, without end.
>
> (1962: 51–52)

Smell can evoke rich memories, but it can also evoke tantalisingly incomplete memories. Whilst place-specific memories are common, olfaction can also give us more abstract memories of kinds of relationship and emotion, but lacking in specific location or point in time. Chesterton, seeking to capture the potential of olfactory experience, uses strongly emotional language typical of novelists and poets seeking to share a specific smell with the reader: 'the brilliant smell of water, the brave smell of stone, the smell of dew and thunder, the old bones buried under' (1958: 163).

Olfaction gives us not just a sensuous geography of places and spatial relationships, but also an emotional one of love and hate, pain and joy, attachment and alienation. Smell can have an influence on levels of anxiety and stress (Arnot 1991) and give happiness and so relieve stress (Birchall 1990). It gives us a geography of life and death. Ileana Malancioiu, a Romanian poet, writes in her poem *The Bear*: 'sniff me well without a sound to see that I am still alive' (1985: 13). And elsewhere Malancioiu uses the word 'putrid' in both its olfactory sense and in a graphically visual sense of decay and death:

> It's raining in a god-forsaken village
> the more putrid trees crumble down
> Putrid with mud is the lane
> only her shining body of a mountain woman

doesn't seem to have become putrid with rain.
She has come to an old cottage
she climbs the steps, goes out of sight.
A putrid window lets in
all that is putrid outside.

(1985: 181)

This is an emotionally charged place evocation, the 'putrid' trees wet with 'putrid' rain, and a 'putrid' window and 'putrid' outside, is one of pain and disgust, of death and misery. This is a poetry from a Romania still under the yoke of Ceausescu's communism, one still poor and repressed.

Porteous notes how travellers and outsiders in general often remember rich smellscapes of the places visited for the first time. These smell memories can be both passionately positive and strongly negative, often associating particular smells with things such as the local cuisine, health and hygiene, history and a sense of the past, the climate and even the feelings of the traveller themselves about their own state as outsider. Outsiders often provide fascinating insights not only into the places visited but more importantly the travellers themselves and the culture from which they come. Early travellers to the Western Isles, for instance, often reported quite negatively on the hygiene of the local inhabitants with ready reference to the odours that met them on entering people's homes and lodging houses (e.g. Cooper 1979). The descriptions of the outsider are often comparative, that is making reference to smellscapes and practices of home and of other places visited on the journey. Particular regions, such as Africa, India, China and the West Indies, have gained quite a reputation for their distinctive odours.

No account of India, from Kipling to the recent popular novels of M.M. Kaye and the accounts of Geoffrey Moorhouse, fails to evoke the peculiar smell of that subcontinent, half corrupt, half aromatic, a mixture of dung, sweat, heat, dust, rotting vegetation and spices. The intimate relationship between smell and the exotic ...

(Porteous 1985: 362)

On fieldwork to North Africa, a European student is immediately struck by the distinctive olfactory geography of the souks of the older quarters of Arab cities. How ugly do the petroleum rich odours of the supermarket car park and city street of the West seem in contrast!

SMELL, PLACE AND CULTURE

A recent survey conducted by the *National Geographic* found that there are significant differences in olfactory sensitivity between individuals within its North American readership and, from its survey in other markets, also found variations in sensitivity to specific odours in different regions (Gilbert and

Table 5.2 Smell ability around the world
% of people who could NOT detect the smell of
androstenone

Country	Male	Female
USA	37.2	29.5
Latin America	24.6	17.7
Caribbean	29.2	17.5
UK	30.0	20.9
Europe	24.1	15.8
Africa	21.6	14.7
Asia	25.5	17.2
Australia	24.2	17.9

Source: National Geographic (1987)

Wysocki 1987). This was only a reader survey and inevitably tended to sample predominantly white middle-class respondents who were American or of European origin, and who, presumably, as a common readership, shared some of the same perceptual attitudes (or culture). Results from this survey seem to suggest that there are significant differences in olfactory sensitivity between males and females, and between different regions of the world (see Table 5.2). Most responses were, perhaps inevitably, from the North American home market and Western Europe but the regionalisation does give an approximate assessment of cultural differences in olfactory sensitivity at the current time.

There is much anecdotal evidence for cultural variation in olfactory sensitivity and for differences in olfactory geographies. It is widely assumed that odour perception is not important to people in North American and European culture, but this idea is 'a relatively recent idea, loosely defined and not based on objective analysis of behaviour' (Engen 1982: 2).

The emergence of urban-industrial societies seems to be associated with major changes in sensuous experience and the definition of the senses (Schafer 1977; McLuhan 1962). Olfactory perception and the role of smells in geographical experience has clearly changed over time as well as differing between cultures at any one time. This is rooted in both cultural practices and the technological conditions of that society. In Western cultures, it is commonly assumed that the nose has become an atrophied organ (Lefebvre 1991). This line of reasoning concludes that olfaction has been marginalised, reduced to a secondary and subservient sense, under the dominance of other senses, notably sight and hearing. This, however, underestimates the extent and significance of olfactory experience in everyday life in both social and environmental relationships. There is a significant olfactory geography in Western cultures which is quite different in nature to the other sensuous geographies and, more importantly, is clearly different from that observed in other cultures today and in the past, including the West itself. It is more

75

accurate to argue that olfactory geographies continually change over time and differ from place to place as cultures do, having both a history and a geography.

Specific cross-cultural studies of olfactory geographies are rare but there are a number of more general studies of cultural difference and specific cultures, such as traditional Arab and Berber cultures and various aboriginal cultures (Hall 1969; Kern 1974; Bourdieu 1977; Lopez 1986; Weissling 1991; Rundstrom 1990). Exploring these differences should not lead us merely to degrade our own olfactory experience but should heighten the understanding our own culturally defined olfactory practices (or experiences).

Nevertheless, to some extent Lefebvre's observation does match contemporary Western experience when compared with other cultures. In West European and North American societies there is a tendency to dislike strong odours, to associate lack of odour with cleanliness of spaces and people, and to prefer simpler body perfumes rather than rich aromas (Gilbert and Wysocki 1987). However, it can equally be observed how the smell of freshly ground coffee adds an essential ingredient to a restaurant or of freshly cut flowers to a living room. Montcrieff (1966) identifies a general preference for natural odours over chemical and synthetic ones and suggests that this is partly explained by a preference for more complex odours (such as raspberry, with over twenty-five constituents), but that the higher the concentration of an odour the less pleasant people in Western cultures tend to find it. This may tally with other culturally expressed preferences, such as for countryside over city, and the relatively open landscape over the enclosed spaces of thick forests and urban alleyways (see Appleton 1975). The tendency for us to interpret words such as 'smell' and 'odour' as negative first and only subsequently to recognise positive olfactory experiences, perhaps also supports the view that in Western culture smells are not so highly valued or enjoyed as in other cultures. However, this is not to deny that we do not greatly value certain smells.

In cultures which maintain a close relationship to their natural environment and in contact cultures where interpersonal relationships are more tactile (see Chapter 4), olfaction seems to be defined quite differently. The people of hunter-gatherer, aboriginal communities, for instance, seem to have had a particularly acute sense of smell, using their noses in following a trail whilst hunting or gathering food in a way not dissimilar to animals. In addition, they often show a greater tolerance for odours in their living environment, whether it is from food (such as fish), materials used to construct buildings (such as skins and cattle dung) or from human body odours (through less frequent washing) (Lowenstein 1966). Early European visitors to Inuit igloos were in particular struck by what seemed to them the filthy smell of these warm and pungent abodes (Weissling 1991; Carpenter 1973).

The traditional Berber dwelling offers a quite radical contrast to the average Western-style dwelling (Bourdieu 1977). Like many pre-modern houses, this structure combines family living space and accommodation for the livestock and stores of food. Bourdieu's primary interest is in the symbolism of the Berber home, but in describing the structure of the house and the meanings attached to these forms he shows much insight into its sensuous geography.

He concisely describes the traditional dwelling's construction and organisation:

> The house is rectangular in shape and is divided into two parts approximately a third of the way along its length by a lattice work wall half as high as the house. Of these two parts, the larger is approximately 50 centimetres higher than the other and is covered over by a layer of black clay and cow dung which the women polish with a stone; this part is reserved for human use. The smaller part is paved with flagstones and is occupied by the animals.
>
> (1977: 98).

What becomes immediately obvious from this description is that this is a co-habitation house, one where all the family and its animals occupy a common internal space; the lattice half wall is only a loose demarcation and does not shield the residents from one another's auditory and olfactory presences. The dividing wall is a visual and tactile shield only. Furthermore, the house is constructed of local materials, many of which have characteristic rustic odours to Western noses. On fieldwork in the Atlas Mountains of Morocco, I was privileged to visit a number of these traditional dwellings. The immediate impression is far less 'organised' than Bourdieu's tidy description implies. One's immediate impression is of dirt and dust, hens running about and a goat perhaps, a lack of furniture of the kind found in Western houses, and an overall compactness of the shared living space. The focus of the house is a courtyard space in the centre which is open to the sky surrounded by cooking space, sleeping spaces, stable areas and stores. The overall effect is 'earthy' and rich with the odours of animals and the produce stored in the house, though this is most obvious in the alcoves and corners of the house. The house is regularly swept by the woman and everything seems to have its place, but this is not the squeaky clean and rather sterile space of the houseproud English family. The Berber dwelling is a sensuous living space.

Comparisons can be made with traditional house forms in Europe, most of which have long since been replaced by modern structures. The traditional croft house in the highlands and islands of Scotland – the *taigh dubh* or black house – combined quarters for the family at one end and for livestock at the other, under the same roof and in the same room. It was a single-room dwelling with thick walls, originally with no windows but having an open door in the middle of one wall, a sod roof and a single opening above the fire to let the

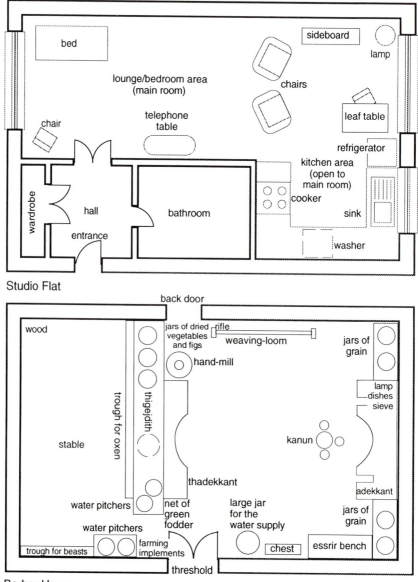

Figure 5.3 The Berber house and the studio flat
Source: Berber house from Bourdieu (1977: 107)

smoke escape. The floor was dirt and covered with straw (Murray 1973).

The Berber house, therefore, is in total contrast with the modern Western house – of whatever size. Western houses are filled with carpet, curtains and furniture (see Figure 5.3). They are totally enclosed spaces, with doors and windows and a roof. The garden is a separate space to the house. The house is awash with electrical gadgets. Generally, there is more space for everyone. Even in a studio flat (one-room bed-sit), which contains kitchen area and lounge in the same room, doubling as a bedroom at night, the bathroom/toilet is a separate room (often with a fan to remove odours and steam), and animals do not share this space. In such a small place, often not even a small pet is kept, and it is common for such flats to be the home of a single person.

In larger Western houses, there are several rooms, each devoted to a specific function and each with solid, full walls and doors which can be closed – kitchen, bathroom, lounge, dining room, bedroom and, of course, the garage for the car. A Berber would have his chickens and other livestock about the house, but we would not even have our car or motorcycle parked in the lounge! Even though many Western houses are in practice untidy and cluttered – not like the neat ideal of a show-house – they are primarily visually organised. Pictures adorn the walls, colours are carefully chosen for the furnishings and decoration. The house, if near a busy road, may be as much double-glazed for sound insulation as for warmth. The windows in modern houses are large not so much for letting in light – the electric light can provide that – but for the view out (see Tuan 1974). The texture of surfaces is quite varied, from lush pile carpets to smooth kitchen tops. Yet, the smellscape of the house is often, to the resident, either largely unconsidered or an embarrassment. The resident tries to minimise odours. The sleeping, cooking and bathroom spaces are quite separate, there is an extractor fan in the bathroom and/or the kitchen to assist the removal of smells, flowers and/or an artificial spray will be used to 'freshen' the air. The horror of every home owner, it seems, is for a visitor to call in when the house still smells of the stale odours of yesterday's boiled cabbage, or for a guest to come down from the bedroom to the lounge next morning before the owner has cleared the air of the hanging stench of cigarette smoke and alcoholic drinks from the night before.

Hall distinguishes between contact and non-contact cultures in attitudes to smells, especially in interpersonal relationships (1969). He contrasts Arab and American behaviour towards smells. Much of Arab society has a long tradition of urban living and the people often live at densities much higher than those experienced in the West. For Arabs, smell is a more prominent sense and is particularly important in social relationships. Enjoying another's smell and permitting one's body odour to greet another person, is a vital part of interpersonal relationships. Disguising one's odour or keeping one's distance beyond the reach of olfactory exchange is suggestive of distrust (Hall 1969).

Table 5.3 Cultural stereotypes of olfactory geographies

Contemporary Western culture	Traditional Arab culture
Negative attitudes	Positive attitudes
Low intensity smells	High intensity or pungent smells
Low tolerance of body smells and many environmental odours	High tolerance to and importance of body smells and environmental odours
Smell and disease association	Smell and trusting association
Odourlessness and hygiene relationship	Health and happiness indicated by rich odours
Synthetic masking and manufacture of deodorants	Natural perfumes, incense and enhancers of odours

(a highly generalised list of themes for reflection)

Further, in public spaces in the older part of traditional Arab cities – such as Marrakech, Fes, Kairouan and in particular the souk market areas of Arab cities – have far richer positive smellscapes than the modern Western city. The streets of cities such as New York and London can often have quite a negative smellscape of traffic fumes, whilst the enclosed shopping malls and department stores are a mixture of zones of relative odourlessness and areas of pungent 'staged' smellscapes, such as the perfume of the beauty products counters. The traditional Arab souk, an maze of narrow passageways, small openings and stalls and workshops has a far richer odour of the products sold – from live animals to leather and cloth – and these smells seem to mingle with great abandon. The specialisation of different areas of a souk for different trades, the confined nature of the souk maze of passageways and the relatively warm climate perhaps make the smells more pungent. The traditional open air or enclosed markets still found in Europe, and in particular fruit markets and fish markets, preserve something of this rich olfactory geography. Yet, sadly, modern packaging can hide much of the olfactory potential of our markets. The supermarket seems a desert when compared to old-style markets and souks where a positive enjoyments of smells is taken into account.

However, in private spaces, such as the house, it is not uncommon for Arabs to burn incense to add a pleasant background smell. This is not the synthetic odour of a deodoriser sprayed to hide or mask stale or bad odours, as practised in buildings in Western Europe and America. Rather, it is a positive addition of a natural odour, released by the burning, to give a pleasant ambience to the living space. The incense is, perhaps, the living space equivalent of perfumes for the body but it is not a deodorant. Arabs have a far more positive disposition towards the smells around them than people tend to have in Western cultures. The traditional Arab house opens to the sky and thus there is a flow of air through the building which ensures that stale odours do not linger or that smells, such as of incense or from cooking, do not build up to excess levels.

Therefore, in the Arab olfactory experience there is a more positive propensity to participate in smellscapes and to utilise them to gain important information about people and places and to communicate to others information about one's products to sell, one's trustworthiness and so on. In Western cultures, there is a tendency towards treating olfaction as a secondary or supplementary experience and to favour deodorisation and a general reduction of smell intensities (with exceptions, such as freshly ground coffee). In contrast, more often Arabs will embrace smells as a vital part of their sensuous encounter with the world, identifying it as important in the establishment of relationships of trust with friends and business partners, as an indicator of health and wealth, and as a significant component in establishing a sense of place and feelings of belonging to particular places and communities.

This alternative olfactory geography is grounded in a more positive attitude towards smell. In spatial terms it implicitly enables a distinction to be made between inside and outside places, both physically and emotionally, and it permits more detailed evaluations of people and places in terms such as trustworthiness, friendliness, wealth and prospects. This geography is sustained by the continued predominance of natural smells – from humans and animals, the traditional trades and from natural perfumes and incense. This more traditional geography is, however, fast disappearing as Arabs adopt non-traditional attitudes and utilise products from Western cultures. The ubiquitous petrol and diesel powered cars and lorries introduce a new olfactory experience to the Arab city, one far less appealing to traditional tastes. Fashion encourages the use of Western style perfumes and deodorisers, both on the body and in the home, many of which are synthesised compounds. Further, many of the consumer products now available for the more affluent residents, such as soaps and detergents, contain both synthesised and non-local odours – notably the 'pine fresh' and 'floral bouquet' which derive from or mimic odours of temperate climates and not of North Africa or the Arabian peninsula. The modern city, which often stands side by side with the older quarters of Arab cities may – apart from the distinctive smells of Arab cooking – smell not dissimilar to those of other modern Western style cities. Olfactory geographies change and such changes are part of socio-economic and cultural changes.

6

AUDITORY GEOGRAPHIES

INTRODUCTION: ORAL, ACOUSTIC OR AUDITORY?

Despite the importance of auditory experience in our everyday lives, we are commonly described as a literary rather than an oral culture, primarily visual rather than auditory (Ong 1971). One can readily observe how our language and thinking is constrained and even distorted by an apparent hegemony of the visual in Western culture (see Chapter 7). Our language, at least, seems dominated by the visual on two levels:

1 *metaphors* – there is a rich vocabulary of words which are used to describe visual experience but are also widely and frequently used metaphorically (and often we have forgotten that they are metaphors and the significance of this fact!) – view, scene, outlook, perspective, prospect, appearance, look, foresight, hindsight, insight, image, etc. (Pocock 1981a); and

2 *structures* – integral to the metaphorical use of visual words is a consequent change in the way we relate to the environment around us and structure our thinking – which is replicated in the structure of our language. We seem to think primarily in terms of synthetic wholes – the view, image, scene – or attend to a particular angle – or point of view, perspective, outlook – and identify discrete objects set in the context of other objects as a relatively stable image. An auditory world unfolds like a tune, a visual world is presented already complete like a painting. The object–subject structure of our language emphasises a relatively static view of the world (Bohm 1983), one derived from a visual interpretation. Bohm considers quantum physics and the difficulty of describing dynamic phenomena in English. He argues that our language is controlled by the noun rather than the verb and he calls for a rheomodal language which is verb-oriented. Closer examination would perhaps question this claim. Verbs play a key role in our language – as is shown by how we readily invent new verbs, often derived from nouns. Yet, Bohm has a point since we seem to see objects and patterns first and

processes and flows second, unlike the Inuit (see Carpenter 1973) who seem to have perceived processes before objects, a flow rather than a scene.

Becker (quoted in Blauert 1983) reminds us that our concepts and descriptions are based primarily on visual objects, for instance: 'the bell sounds' and not 'the sound bells'. However, we should not overemphasise this visualism in Western culture. It is true that visual culture and technology (from painting to photography and film, from microscopes and telescopes to fibre optics) is well developed in contrast to the olfactory and tactile realm, but the auditory is also very important in our culture and technology (from music and language to amplifiers and telephones). We must not overemphasise the visualism of our language nor underestimate our reliance on the ears.

The distinction between literary and oral cultures needs some clarification. This primarily refers to a distinction between cultures which preserve their history by written records and utilise reading and writing in everyday life, and those which preserve their history in stories remembered and recited by its people as a spoken history, like oral history, and generally not relying to any great degree on reading or writing in everyday life. It is often observed that oral cultures are also more sensitive to the sounds in their environment (such as the aboriginal Inuit and Saami) and it is assumed that literary cultures are somehow inferior in their auditory capacities. This is, however, far from accurate. Both oral and literary cultures appear to have highly developed auditory sensitivities. For instance, music is a highly developed art form in both aboriginal oral cultures and modern Western societies – though it is often quite different in style. In literary cultures, techniques have been developed to write down the music – notation – whilst in oral cultures musicians pass on the artistic tradition with the instruments, just as the people pass on their history with their voices. The key distinction for auditory geography is perhaps not between literary and oral cultures – since this confuses the issue – but between 'oral' and 'aural' cultures. 'Oral' refers to the voice, whilst 'aural' (or auditory) refers to the ear. Cultural differences in approaches to the auditory dimension do appear to be significant, though it is difficult to argue that some cultures are more or less auditory – rather they define it differently.

Interest in sound in the environment and the ears in geography has grown in recent years but there is little agreement over the correct label for the field (Pocock 1987, 1988, 1989; Porteous and Mastin 1985; Kariel 1980, 1990; Ohlson 1976; Southworth 1969). Oral geography would refer, presumably, to a geographical knowledge spoken in the same way as oral history is related. Schafer (1985) prefers the term 'acoustic space' but this most strictly seems to suggest an interest in the space created by sounds, such as the range of a church bell or the echo in a building. Schafer (1977) has also used the term 'soundscape' to describe the sonic environment surrounding the sentient, drawing an analogy to 'landscape'. This term has been adopted by geographers (Porteous and Mastin 1985). Aural geography would represent a sensuous geography derived from the ears, though the sound of this word is so like 'oral' that despite the neat play of meanings its spoken use could generate – linking hearing and speaking – it is avoided. Pocock (1988) refers to the 'music of geography' which encourages a positive attentiveness to the auditory experience, but can also be confused with music geography, the study of the geography of musical traditions and the relationship of music and environmental sounds (Nash 1986). Sonic geography would refer to the spatial organisation of sounds and characteristics of places in terms of sound. Focusing on the sensuous experience, we might refer to hearing or listening, but which would we choose? Hearing represents a passive experience, listening an active one. The term 'auditory' can encompass both.

In this chapter, the term *auditory geography* is chosen since it relates specifically to the sensuous experience of sounds in the environment and the acoustic properties of that environment through the employment of the auditory perceptual system. The ear forms the main sense organ in this perceptual system but, as we have already observed in previous chapters, geographical experience is multisensual and ecological, that is we can 'hear' with more than our ears and the context, or environment itself, plays a key role in what or how we hear.

SOUNDSCAPE

Precise description, either through technical terms and/or formal equations and models, is an important part of developing an accurate understanding of a phenomenon and enables us to share that understanding with others. Raymond Williams perceptively pointed out that 'we learn to see a thing by learning to describe it' (1965: 39). Schafer (1977) and the World Soundscapes Project offers a standardised vocabulary which is designed to assist the study of auditory experience in different places and times. It is widely used but often quite uncritically. This terminology has been translated into a number of languages with varying degrees of success. The French *le paysage sonore* provides a good equivalent to 'soundscape'. However, in German, various terms have been tried with limited success (see Schafer 1985: note 4). The key merit of Schafer's vocabulary is simplicity. This is also its limitation. It combines a redefinition of existing terms borrowed from various sources and the invention of new terms. Though Schafer himself is a musician and composer, his terminology is heavily reliant on visual metaphors.

A key distinction is made between 'soundscape' and 'soundfield'. The 'soundfield' is the acoustic space generated by the sound source, that is, the area spreading out from the sounding or voicing agent. The soundfield of a tolling bell extends out into the countryside or across the town. The loudness of the sound, or the power of the soundfield, will generally diminish with

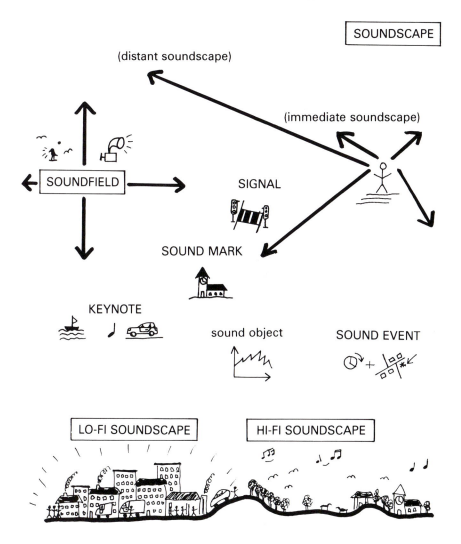

Figure 6.1 The Schafer terminology

distance from the source. The emitter, or voice, is at the centre of the soundfield not the receiver or sentient. The soundfield is generally charac- terised by a single sound. Many different soundfields may overlap across a given space. This overlapping of a multiplicity of sounds from different sources, and probably from different locations, produces the soundscape. (Blauert (1983: 2) distinguishes between 'sound event' and 'auditory event'. These are similar to soundfield and soundscape but must not be confused with them. The 'sound event' is a physical aspect of hearing and relates to the source. There is no geographical connotation here and I avoid the term because of potential confusion with Schafer's own term 'sound event'. Blauert's 'auditory event' refers to what is perceived, but again excludes the wider geographical aspect of soundscape.)

The 'soundscape' is the sonic environment which surrounds the sentient. Soundscape is shorthand for 'anthropocentric sonic environment' (Ohlson 1976). The hearer, or listener, is at the centre of the soundscape. It is a context, it surrounds and it generally consists of many sounds coming from different directions and of differing characteristics. It is the sonic equivalent of landscape. Schafer defines soundscape somewhat vaguely as 'any portion of the sonic environment regarded as a field for study. The term may refer to actual environments, or abstract constructions such as a musical composition or tape montages, particularly when considered as an environment' (1977: 274–275).

The term 'soundscape' is deceptively attractive. In English, this '-scape' suffix occurs in a range of words, like the suffix '-ship' which seems to suggest a state of being, as in friendship, kinship and relationship (Relph 1981: 26). In Schafer's use of the term 'soundscape' there is an implicit parallel made with the widely used term 'landscape'. This originally referred to a unit of land held by a group or individual and, by association, a relationship to that land, and this term was later adopted and came to almost exclusively refer to a style of representing the world around us in painting (see Chapter 7). In this sense, 'soundscape' seems to suggest a kind of state of being with respect to sound in the environment and/or the depiction of the world in terms of sound like a sound picture or landscape. This implicitly links the soundscape concept to traditions such as painting and architecture, and ideas of linear perspective and the composed view or scene. In this visual tradition, the observer is detached, the image is static and is viewed from a privileged position. Auditory experience is far more dynamic and the sentient partici- pates within the sonic environment. There is a tension within Schafer's use of the term. On the one hand, and more often, he uses 'soundscape' to refer to a geographical space of particular sonic characteristics. This is generally the definition adopted by others using the term (e.g. Porteous and Mastin 1985). This suggests soundscape as an aesthetic object, not unlike the painting or musical composition (the latter being one of Schafer's own analogies). However, there is a second use of the term which is generally neglected:

soundscapes as auditory experience. Here it is less an object for contemplation and more a process of engagement with the environment. The soundscape moves with the sentients as they move through the environment and it continually changes with our behavioural interactions. In this sense, one cannot 'map' a neighbourhood soundscape – to do so is to suggest a kind of soundscape as object (as in Porteous and Mastin 1985). Soundscapes surround and unfold in complex symphonies or cacophonies of sound. Using the term 'soundscape', we must remember these qualities and not allow visual connotations to usurp our understanding.

Schafer also recognised a smaller unit: the 'sound event' is the smallest self-contained particle of a soundscape which can be defined by the human ear. It is a particularly useful term since 'event' refers to something which occurs in a certain place during a particular interval of time. All sound is within a time–space continuum, a series of events, and the sound event beautifully describes the dynamic nature of auditory experience. The view is a stable image (even when movement is observed), but the auditory experience is a flow of sound(s), constant for awhile or rising or falling in intensity (loudness) or pitch.

These terms can be refined. Ohlson (1976) and Grano (1929) distinguish between an immediate soundscape and a distant soundscape. The 'immediate soundscape' corresponds to the visual landscape and the close correlation of the two senses in the geographical experience of the near-environment. The 'distant soundscape' is more completely auditory and here the identity of the sound source is more ambiguous or hidden by being beyond what can be visually confirmed. Schafer also distinguishes foreground and background qualities but uses terms derived from painting and music related to the figure–ground framework. 'Keynotes' are recurring, generally repetitive, background sounds against which the other sounds in a soundscape may be heard. In music a keynote is the key or tonality of a particular composition. It could also be used to describe a ground-base drone. The keynote provides the fundamental tone around which the composition may modulate and from which other tonalities take on special relationships. 'Keynote', therefore, is an extremely useful concept. In a soundscape the keynote is heard often enough in a particular society to form a background against which other sounds are perceived. Schafer suggests that examples might be the sound of the sea for a maritime culture, or the sound of the internal combustion engine for the modern city. Keynote sounds are often not consciously perceived, yet they condition or influence the perception of other sound signals. 'Signal', or 'sound signal', is defined as figure or foreground: 'any sound to which the attention is particularly directed' (Schafer 1977: 275). The signal is generally a readily identified sound, an informative sound which stands out above the overall soundscape texture. It is important to distinguish it from a similar term, 'soundmark'. 'Soundmark' derives from the term landmark. It refers to a community sound, that is a sound recognised and shared in the soundscape of a social group. It is a relatively unique sound or specific to a particular community, and it possesses qualities which make it special or noticed by people in that community. Often this marking of the sound is a combination of the character of the sound, such as the sound of a bell or hooter, and the associations of a tradition, such as bells and religion, hooters and factories. Soundmarks may therefore have important symbolic qualities.

For the auditory experience itself, Schafer (1977) distinguishes between 'lo-fi' and 'hi-fi' soundscapes. The 'lo-fi soundscape' has an unfavourable signal to noise ratio. It has so much sonic information that little of it is heard distinctly. This is more typical of the city than the countryside. The discrete sounds are jammed by the general urban-industrial background. Heavy traffic in contemporary cities is a major contributor to lo-fi street sound-scapes where the details of bird song and wind in the trees is submerged in the drone and thunder of the internal combustion engine. A mighty waterfall, such as Niagara Falls, offers an example from nature. 'Hi-fi soundscapes' have a favourable signal to noise ratio. There is a low ambient noise level and discrete sounds emerge with clarity and may show interesting relationships. The hi-fi soundscape is generally a feature of the countryside and seems to have been a feature of pre-industrial geography.

When describing the sound of machines as noise rather than music, we are perhaps in part distinguishing between lo-fi and hi-fi soundscapes. In some cases, it is not necessarily the intensity (loudness) of ambient sounds as the character of them which effectively jams other sounds in lo-fi soundscapes. A high-pitched, continuous or repetitive sound, for instance. It seems that lo-fi soundscapes are more often induced by human activities. In the hi-fi soundscape, the human voice can always be heard in conversations. In the woodland or a wetland at dawn or dusk there may be much sound from many birds, yet the song of each is discernible and each can communicate with its own kind.

We might equate lo-fi soundscapes with cacophony and hi-fi soundscapes with symphony. However, unlike the symphony in music, these soundscapes in the environment – whether rural or urban – are not compositions as such but accidents of evolution. When we try to identify more precise composed elements, such as the melody, rhythm or key of a soundscape, we are composing the sounds ourselves, that is, making sense of what we hear. The sense or meaning of sound is therefore both relative to a possible source/ activity and in relation to other sounds. More generally, sense or meaning is derived from previous experience of the same or similar sound and from the context of our experience, activity or intention. The environment itself also structures what we hear (see Chapter 2). However, the lo-fi/hi-fi distinction does lack precision. Whilst examples of each can be identified, most auditory experience is more ambiguous and fluctuates between degrees of cacophony and symphony. How should signal and noise be defined – physically or culturally? Different individuals might discern quite different sonic qualities in their environments. Parallels can be observed in olfaction: a negative smellscape – one with an overall disliked stink which we are reluctant to explore is 'lo-fi'; a positive smellscape – one with a rich aroma of pleasant smells, invites olfactory exploration and is 'hi-fi' (see Chapter 5).

Soundscape, and the associated terms, provide useful words for an auditory geography and identify important dimensions for investigation. However, its metaphorical basis reminds us of the limitations of language, the complex relationships between the different senses – especially between sight and hearing – and the need to consider here the specifically auditory qualities of sensuous geography.

SOUND, SPACE AND PLACE

Sight and hearing are commonly described as distant senses (Hall 1969; Gold 1980; Tuan 1993) and in Western cultures, at least, seem to be particularly important in geographical experience. Closer examination of auditory experience, however, undermines simple distinctions such as intimate versus distant senses and reveals a wide variety of auditory encounters with the environment and complex inter-relationships with the other senses.

The sensuous matrix (see Chapter 3) provides a useful initial classification of the different aspects of auditory geography. The four elements identified are often confused but each distinctively contributes to geographical experience (see Figure 6.2). Hearing may be described as the basic passive sensation, whilst listening implies an active attentiveness to auditory information and a desire to establish meaning. Furthermore, not only is auditory information acquired about a world, an environment of sounding things and organisms in inter-relationship, but also things and people emit sounds, or

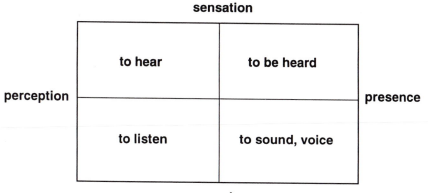

Figure 6.2 The auditory matrix

have a voice, which projects them into that world. Expressed alternatively, we not only perceive of a world, but have a presence in it (Ihde 1976). The voice is not the only sound emitted by the individual which spreads across space, but also the sound of the body moving through the environment or against it. However, auditory experience is not reciprocal in the sense of haptic experience, for hearing or listening and sounding or giving voice can be quite independent activities. In the present chapter, we will concentrate on listening rather than voice; perceived geographies rather than projected ones. Both sounds received and sounds made literally take place or have location, and occupy or project over space *and* each sound has a specific duration, so occupies time as well. Auditory geography is therefore time–space geography, a dynamic geography of events rather than images, or activity rather than scene.

The auditory sense is primarily physical rather than chemical. Sound is concerned with motion or activity, that is with vibration and resonance of substances (air, water, solids). The ear is a finely tuned mechanism for recording vibrations transmitted through air. Gibson (1968) describes it as an

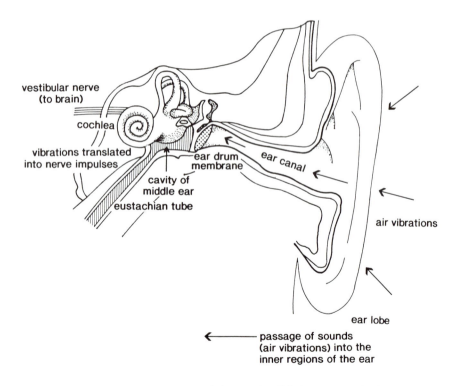

Figure 6.3 The structure of the ear (simplified representation)
Source: Gibson (1968: 76)

auditory system and argues that the perception of sound involves listening not just hearing. This listening system includes two ears on either side of the head together with the muscles orienting them to the source of the sound. The ears collect sounds and through the movement of the head acquire information on the direction of those sounds. The auditory system absorbs information about the character of sounds, their intensity and pitch for instance, and their direction and duration. It not only involves the receptors in the ear (Figure 6.3) and their sensitivity, coordinated with movement of the head (and body), but also the evaluation of auditory information is always in reference or relationship to the body. The ear is the focus of our auditory perception, but not exclusively so. The ear is the most direct of our auditory sensors but the body can also receive auditory information, especially from solid substances. The ear is a complex organ dedicated to collecting vibrations from the air and transforming them into nerve impulses which are interpreted by the brain (Figure 6.3).

The body has its own auditory presence, both explicitly through the vocal chords and implicitly in the friction of its movement (internally and against the external environment) and, most importantly, it has its own biorhythms which allow us to measure the pattern of sounds (rhythm, pace, duration). Auditory perception is against this corporeal background and in reference to it. We become most aware of it when we suffer disablement, are in an extremely quiet place or have learnt to meditate upon the body in itself. Therefore, auditory perception is like all other forms of sensuous experience in that it involves the whole body whilst at the same time giving the immediate impression of sensing from a particular point or dedicated organ – touch and the finger, seeing and the eye, smell and the nose, hearing and the ear. Sensuous experience, even of a particular sense character, is always inclusive of a wider body involvement.

Gibson (1968: 75) makes a distinction between two functions of the auditory system:

1   *exteroceptive* – this picks up the direction of the sound event, permitting orientation to it, and the nature of the event, permitting identification of it;

2   *proprioceptive* – this registers the sounds made by the individual, especially in vocalising. Hearing one's own voice permits control of sequential patterns of vocal sounds, as in bird songs, and the monitoring of social utterances, as in human speech. The ability to speak, as opposed to just making sounds, is made considerably easier by an ability to clearly discern one's voice.

In sensuous experience, the auditory world not only surrounds us but we seem to be within it and participants. Although we 'know' that the visual world is all around us, our actual experience is of an image in front of the eyes and not behind our backs. We feel more detached from a visual world than an auditory one. Auditory phenomena penetrate us from all directions at all times. The auditory perspective is not linear but multidirectional – even when we are deaf in one ear. We might try to dominate space with sound, as with the church bell or minaret amplifier, but once it has left its source it leaves our control (except in the special case of the acoustic design of concert halls). The physical form of the environment (open spaces, enclosed volumes, etc), the wind and the relative moisture of the air and even the presence of other sounds – sometimes background sounds of which we are largely unaware – all modify the impact of any given intensity and character of sound emitted. (The term 'character of sound' refers to the range of potential qualities of a sound – intensity, pitch, rhythm, etc. The term 'intensity' is used to refer to the loudness or volume of the sound, its strength. Metaphors abound in descriptions of sound, often with neglected associations, e.g. volume of sound, texture of sound.) The echo lingers in time and space, and is consequently far more complex than the reflection displayed on a surface. (Yet visual reflections are modified by the surface on which they are projected.) Sounds fill spaces and when they are present give character to particular places, as soundscapes. The visual subsists in the presence of light and is dependent on the angle of observation. Therefore, whilst sound is about volumes and motion, the visual is about surfaces and stability.

The ecological theory of perception, as we noted in Part 1, identifies the importance of the environment in structuring sensory information. Gibson (1968) concentrated mainly on visual experience but his ideas are especially pertinent to understanding the experience of soundscapes. Just as the ambient light received by the eye is structured by – that is, contains information about – the surfaces from which it is reflected, and excessively bright direct radiant light gives us little information or gives us an experience of blindness, so also with auditory experience. The structure of the environment, its openness and enclosedness, and the properties of the materials in that environment, sound absorbing or resonating/reflecting, influence both the sort of sound that is actually heard – its intensity or volume and texture or timbre – and the distance over which it might travel. Sounds are emitted from many sources, and are of varying intensities and different characters, and are modified by the environment through which they pass before reaching the ear. Just as certain colours or intensities of light may mask or modify other light sources or reflections, so particular sounds may mask or modify, or combine with, other sound sources and echoes or reverberations. Therefore the wonder of the auditory system, as with all perceptual systems, is the way it manages to decipher an order, a sense of the world, and of people, places and spatial relationships from this complex mass of sensuous information.

Human beings, it is frequently argued, are primarily visually oriented and the world of our other senses is much less developed (Blauert 1983). The present book, perhaps, has begun to question this common assumption. Nevertheless, when compared to other animals our sense organs for hearing and smell seem less sensitive. It is widely recognised that human hearing is not especially acute and its range is far more restricted than many other animals (Milne and Milne 1962; Katsuki 1982). However, we appear to attach a far greater complexity of meanings to the sounds we hear and our ears are finely tuned to the requirements of spoken language. This apparent compensation of quality for quantity of hearing is perhaps the result of our more highly developed mental skills. Certainly, musicians and composers who have trained their hearing often demonstrate more acute hearing than the untrained. Also, the inter-relationship between our ability to speak and our hearing is often overlooked in the literature, except in accounts of the deaf where the relationship is most obvious (e.g. Mindel and Vernon 1987). In the wider literature, the more common reference is to biological evolution and it is claimed that this has progressively reduced the necessity of acute hearing and olfactory sensitivity. We compensate for our less acute hearing not only by training our ears – or 'ear cleaning' (Schafer 1967) – but also by technological ingenuity and, in particular, various kinds of amplifier and other auditory tools.

The hearing range of an average young human is approximately from 16 Hz to 20 KHz (Blauert 1983: 2). Sensitivity below this range can give the annoyance of being able to hear one's own heartbeat. The human ear discriminates against low frequency sounds and so we do not hear deep body sounds such as the movement of blood in our veins. The upper limit of

Table 6.1 Decibels of common sound intensities

Intensity (dB)*	Sound**
130	4-engined jet aircraft at 120ft
120	threshold of pain; pneumatic hammer at 3ft
110	boilermaker's shop; typical 'rock' band
100	car horn at 15ft; and symphony orchestra playing fortissimo
90	pneumatic drill at 4ft; lorry at 15ft;
80	noisy tube train; loud radio music
75	telephone bell at 10ft
70	very busy London traffic at 10ft
60	conversation at 3ft; a car at 30ft
50	a quiet office
40	residential area with no traffic; and subdued conversation
30	a quiet garden; whispered conversation
20	ticking of a watch when held to the ear and 'silence' of a broadcast studio
10	the rustle of leaves
0	threshold of audibility

* Over 90 decibels can lead to damage to the human ear.
** Approximations which give only a general impression of relative sound intensities.
Source: Fry (1979: 94)

human hearing is very modest when compared to cats who respond to sounds up to 50,000 or bats at 120,000 cycles per second (Katsuki 1982). There are, however, examples in the literature of extraordinary feats of hearing. For instance, Bombaugh (1961: 280) records a slave on 17 June 1776 hearing the battle of Bunker Hill at a distance of 129 miles by putting his ear to the ground. Ohlson (1976) offers a more geographical description of human hearing. He notes that the 'anthropocentric sonic landscape' reaches as far as 15–20 kilometres from the receiver in open ground. Obviously this is under ideal conditions, when other nearer sounds are not competing or masking these distant sounds. The topography, buildings, wind and climatic effects can all influence this range and even create 'shadows' protected from the sound (or 'passageways' where it is amplified). In the stillness of night, it seems we hear more distant sounds than in the bustle and heat of day. However, Ohlson's figure is the greatest distance over which one might ordinarily hear the sound of thunder.

Our hearing is binaural, that is we have one ear on each side of the head. Unlike the forward facing eyes, the ears – at their most alert – can give us an all-round auditory receptivity. Ohlson (1976) argues that binaural hearing

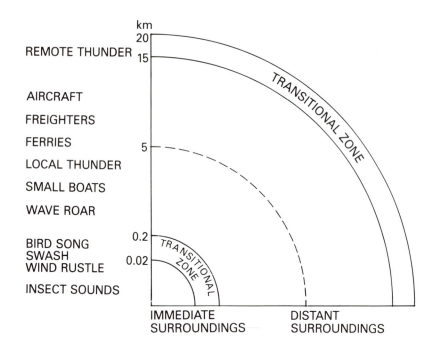

Figure 6.4 Ohlson's sound environment. A general outline of the anthropocentric sonic landscape in the archipelago of south-west Finland
Source: Ohlson (1976: 35)

means it is relatively easy for humans to determine direction of sounds but the auditory sensation gives only an uncertain idea of the distance to the sound source. This argument parallels Gibson's (1968, 1974) concerning stereoscopic vision. He argues that it plays only a limited role in the perception of depth because perception is primarily ecological. Likewise, it can be observed that even a person with only one active ear can still hear the direction of sounds, though perceptual accuracy may be partially impaired on the deaf side. In the immediate soundscape there is a close relationship between visual and auditory perception, the former helping to confirm the latter especially. For the distant soundscape, where sound sources are several kilometres away and mostly invisible, this visual confirmation is not so available.

Age and health affects the physical range of hearing. Although the young may play their music the loudest, the reality is that a healthy youngster will hear fine, high-pitched sounds at some distance, such as grasshoppers on a hillside. The shift of hearing thresholds with age means that this same sound can be hardly heard by men over 60 and women over 65 years old. The old not only begin to have to turn the volume up on their television sets but also start to lose their ability to decipher complex mixes of auditory phenomena, such as conversations on the television from the noises within the room in which they are sitting. This sensory loss plays an important part in the emergence of distinctive experiential geographies of old age (Rowles 1976). Furthermore, we all experience a temporary decline in our hearing, both in terms of loudness and discrimination, when exposed to extreme intensities of sound – such as in a factory or discotheque – and with the general tiredness of a long day. For example, one might play the radio at a lower volume at the start of the day than at the end, enjoying the dawn chorus of the birds, but finding the noise of late evening traffic muffled by the tiredness of the ears after a long day and needing to turn the radio volume up. Sensuous geographies therefore change over time, through a day and through a lifetime.

However, sound is not just sensation: it is information. We do not merely hear, we listen. Hearing is the first sense, and the last to depart when we go under the anaesthetic (Pocock 1988). The human infant can make distinctions between pleasant, soothing and disturbing sounds long before it can discriminate visually with equal subtlety. The child can begin to master the complex process of communication and the rudiments of spoken language long before it has acquired an ability to fully understand perspective and visual arrangement, especially of distant objects, and this visual 'backwardness' persists for some considerable time in children's drawing. Our experience of sound is not merely physical. It is emotional. Because we cannot close our ears as we can our eyes, we are more vulnerable to sound (Wyburn et al. 1964). 'Music is for most people a stronger emotional experience than looking at pictures or scenery', claims Tuan (1974: 8). Music, or musak, can play an important role in selling products or creating an appropriate ambience in public spaces – the band in the town square, the music in restaurants and the take-off and landing music on aeroplanes.

The importance of hearing is immediately evident in the experience of sudden deafness. An acute sense of loss is experienced and deep depression, loneliness and paranoia can develop (Knapp 1948; Rayson 1987). Yet we are also able to imagine, or dream, auditory experiences, like the composer thinking through his composition mentally and the deaf Beethoven conducting his new symphony with great frustration at the mismatch between his hearing-in-the-mind and the visually apparent playing of the ensemble. Withdrawal is a common social response for individuals experiencing gradual deafness and society may appear to exclude them (Lane 1988). The auditory dimension is therefore important to social as well as individual geographical experience.

Ong (1971) notes that the world of sound is an event-world whilst vision is an object-world. Schafer suggests, it '... is a world of activities rather than artefacts, and whenever one tries to write about sound or tries to graph it, he departs from its essential reality' (1985: 88). Pocock also argues that sound

> is dynamic: something is happening for sound to exist. It is therefore temporal, continually and perhaps unpredictably coming and going, but it is also powerful, for it signifies existence, generates a sense of life, and is a special sensory key to interiority – unlike sight which presents surfaces.
>
> (1988: 62–63)

Auditory experience – or sound – plays a key role in anticipation, encounter and memory of places.

Ihde (1976: 50–51) notes that 'silence is the horizon of sound ... the invisible is the horizon of sight'. Visual geographies stretch as far as the eye can see and fade with the fading light. Auditory geographies are dependent on activity and things sounding and the ear being attentive to them. Listening may reveal that silence is not so silent but an absolute silence (perhaps only existing in a lifeless vacuum or void) would be without geography, location or spatial arrangement. With both listening and voice we participate in a geography of a living world, an auditory world, and so distinguish places and identify relationships across space.

AUDITORY GEOGRAPHIES

Much insight can be gained into the structure and character of auditory geographies by considering the contrasting geographical experiences of individuals with deafness and others with blindness. The totally deaf individual has a 'quiet ear' (Wright 1990), that is the main organ of hearing, the ear, provides no auditory information, but this does not mean that the deaf have no auditory experience. The 'quiet ear' may be silent but the body 'resonates', that is it can give at least a partial sense of the auditory world because sound is first and foremost vibration. The deaf can therefore teach us something about the role of the body in the auditory perceptual system. By contrast, the blind often find that without the dominant eye, their auditory faculties appear more acute or at least are more central to their sensuous experience. Many blind individuals rely heavily on touch and hearing to navigate about the environment and some even refer to a form of 'echo-location' which can provide extremely accurate indications of the structure and character of the surrounding environment (Hull 1990). In particular, the blind find their auditory capacity strongly influenced by the context of hearing and thus remind us of the important role of the environment in structuring auditory information (see Gibson 1968). The blind remind us of the sheer richness of auditory experience and the extent to which the environment itself structures the sounds reaching our ears.

The totally deaf and the fully blind are two extremes of potential auditory experience. They help us to step outside of the confines of ordinary, taken-for-granted sensuous experience. In particular, they remind us of the importance for auditory geography of the basis of our sensuous experience in and through our bodies, with the other senses, as well as the mind, and the role of the environment within which sounds are experienced.

There is much misunderstanding about the term 'deafness'. To some it implies total inability to hear, to others a failure to understand the spoken word immediately, and others use the euphemism 'hearing impaired' (Mindel and Vernon 1987: xvi). To understand deafness, we therefore need to appreciate what is meant by hearing and the significance of comprehension, that is 'sense' as both sensation and meaning.

As noted earlier, there is a difference between hearing and listening. To hear is to register an auditory sensation or vibration. To listen is to pay attention to auditory phenomena and perhaps identify them, though not always. Even hearing individuals can be 'deaf' or inattentive to the auditory world around them. Through trained listening we can hear more effectively. A mere cacophony or noise can become a delightful music – a symphony of nature or an urban tone poem (see Pocock 1987). It is commonly recognised that ear training enhances the normal hearing person's hearing discernment or accuracy, as in music training (e.g. Schafer 1967) and that recognising a sound (as in language) heightens auditory experience. The incomprehensibility of a foreign tongue reminds us of this. In deafness, one needs to distinguish between *absolute hearing loss* – an inability to hear the loudness of sounds; and *relative hearing loss* – an inability to differentiate sounds. Mindel and Vernon (1987: xvi–xvii) 'have found it most meaningful to define deafness as a loss of hearing sufficiently severe to render an understanding of conversational speech impossible in most situations with or without a hearing aid'. This combines the two dimensions of hearing: the reception of

97

sensations and the comprehension or deciphering of sensory information. Loss of comprehensibility is perhaps most sadly experienced by those suffering hearing loss with increasing old age (Lyson 1978). One cannot distinguish or organise sounds, the sense or meaning of the auditory world eludes the hearer, and so the geographical value of their auditory experience is greatly reduced. A person may therefore be effectively deaf with a noisy ear. The latter is graphically illustrated by the unfortunate experience of many hearing-aid wearers who find that they sometimes become confused by a mixing of foreground and background sounds. They may still have difficulty in understanding face-to-face conversation yet hear someone speaking at the other end of the room! This creates a kind of 'upside down' auditory geography and confuses geographical orientation, making near things seem far and far things seem near. An ability to differentiate sounds is vital in discerning accurately a geographical world. Sound intensity (or volume) alone is insufficient to differentiate the complex structures of the environment, loud things are not always closer and quiet things are not always distant. Furthermore, those with or without hearing loss can suffer from the blanketing effect of particular sounds. Especially in urban-industrial environments, the dominance of certain loud or repetitive sounds may mask weaker or quieter sounds and effectively hide a part of the environment.

The experience of the fully deaf reminds us that one does not only 'hear' with the ears but with the whole body. Similarly, Bartley (1972) argued that 'people see with their whole body', that is, perception is global. The ear is only the most specialised and perhaps most effective auditory tool we possess. The deaf can 'hear' but not with the ears. They hear through the vibrations felt by the body, sometimes called 'touch-hearing' (Wright 1990). For those who once did hear and therefore have some subliminal memory of auditory sensation, the deaf person may also 'see' sound as a kind of 'eye music'. The wings of birds 'sing' or, as Wordsworth observed, 'a soft eye-music of slow-waving boughs' (Wright 1990). However, the still windless day is especially silent to such 'hearing'.

David Wright (1990) provides an interesting account of a man with profound deafness who became deaf in early childhood, but he reminds us of the uniqueness of individual experience: 'about deafness, I know every-thing and nothing' (1990: 5). It is important to distinguish between those born deaf and those individuals who become deaf suddenly or gradually at various stages through life. Those born deaf can have quite serious communication problems, though these can be overcome. Those individuals who become deaf have quite varied experiences depending in part on when and how they became deaf. Wright's experience is largely positive but he is very conscious of the misunderstandings of hearing people. In particular, he finds 'irritating' the theory that the loss of one sense is compensated for by a quickening of another. 'There are no compensations, life is not like that. At best we are offered alternatives. We have no choice but to take them' (Wright 1990: 12).

Interestingly, Wright has had a lifelong passion for poetry and the sound of words, even enjoys trying to say them and feel them in his throat. Despite deafness, he appreciates something of the auditory world. Though he is 'completely without hearing' in the conventional sense, 'I do not live in a world of complete silence. There is no such thing as absolute silence. Coming from one whose aural nerve is extinct, this statement may be taken as authoritative' (Wright 1990: 9). He reminds us that all sound is *vibration* and the ear, roughly speaking, is a highly specialised organ for the reception of air vibration or sound waves. Other materials also conduct vibration. The deaf person can 'hear' footsteps behind them through the feet feeling the

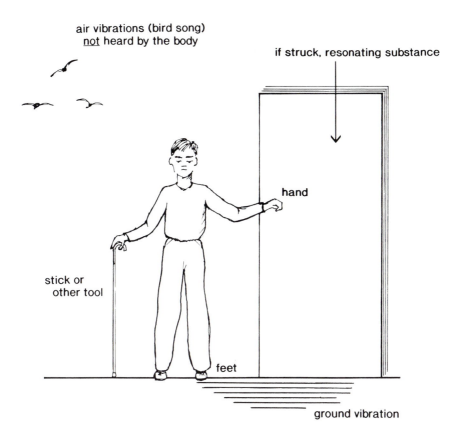

Figure 6.5 Hearing with the body (dependent on the resonating properties of substances *in contact* with the body)

vibrations on the wooden floor. A less resonant substance, such as stone or concrete, may have much echo in the air but little resonance as a material and so appears silent to the deaf. However, though some sounds have presence in this way, it is often quite difficult for the deaf person to locate the source of the sound or vibration since the vibration resonates across or through a zone of the environment or, more importantly, along particular resonating substances in the environment. Hearing with the body is thus quite different from hearing with ears, and consequently can generate a quite different geographical experience. It is perhaps more akin to the geographical experience of touch, especially global touch, and of olfaction, giving more a sense of presence than of precise spatial relationships and locatedness. However, resonating substances or the feeling of vibrations do yield a certain kind of auditory geography.

Wright can lip-read but was taught as a child to vocalise, so that through a combination of speaking and lip-reading he can engage in conversation with hearing people to a certain extent. He can feel his own voice but cannot judge its loudness or quality. All forms of hearing are selective. Body hearing – that is feeling vibrations through the substance of the body and through its structure – is selective of lower frequency sounds. However loud the sound, that perceived by the deaf is incomplete. Conversation and music are largely impossible to make sense of. Something can be 'heard' of stringed instruments and percussion, especially drums, through their resonating properties. Wright lists some of the sounds he can 'hear' – including gunfire, detonation of high explosives, low-flying aircraft, cars backfiring, pneumatic drills and other deep and resonating sounds – but notes that these are generally sensed abruptly without the preparation of sounds which the hearing individual would ordinarily pick up (often as part of the background sounds) but the deaf person cannot. Therefore, the auditory experience of the deaf is restricted to low frequency sounds, those that resonate relatively strongly through solid substances, but is characterised by abrupt appearance and disappearance, that is, discontinuity (though not always). The continuous play of background and foreground sounds is absent for the person listening 'with the body'. In addition, one's own body rhythm, or noise, can muddle environmental perception. The world of the deaf is quiet but seldom silent.

Wright recalls the delusion that out of sight equals out of ear-shot. It is a kind of reflex that as soon as someone passes out of his field of vision, he assumes that *he* is inaudible to them, even if they are a hearing person.

If I am talking to someone and they turn a corner or disappear behind a screen, I involuntarily stop speaking. Even though I know he can hear me, my subconscious won't believe it. Walking with people at night, in the dark, I fall silent; so strong is the reasonless conviction that hearing depends on seeing.

(1990: 110)

For the deaf, with the coming of night the world literally draws in with the darkness. The blind experience a similar type of delusion associated with taken-for-granted sense associations. Hull (1990) recalls how not seeing can easily be translated, by the blind, into the assumption of not being seen. The blind person must always be aware that a seeing world invades their apparent privacy.

The deaf, hearing with or through the body, experience a quite different auditory geography to that provided by the ears but, as we have observed, can remind us of important properties of auditory experience. Whilst the ears provide continuous and complex combinations of background and fore-ground sounds of widely varying pitches, timbres and intensities and often over great distances, the body offers a more limited and selective auditory experience, more intimate and fragmented, a world of vibrations – a kind of touch-hearing geography – of predominantly low frequency sounds and highly dependent upon the resonant properties of surrounding substances, generally lacking discernible timbres and pitches, often abruptly appearing and disappearing, and extremely difficult to associate spatially with particular sources and locations.

The blind tend to rely much upon both their haptic and auditory experience and reflection on their experience of these perceptual systems in navigating in space and relating to places can give us much insight into the potential of our tactile and hearing senses (Lusseryan 1963, 1973; Blackhall 1971; Sullivan and Gill 1975; Hull 1990). For those born blind, the distinction between the senses may be less clear (Hill 1985). The haptic experience provides a continuous body-contact geography, whilst auditory experience provides a more extended or distant geography, an experience of wider spaces and the relationships between places. Further, the sensuous experience of blind people helps us to more clearly appreciate the way in which hearing is not just dependent on the acuity of the ear itself but that which is heard has been shaped by the environment through which the sound has passed. One does not just hear voices and sounds but one listens to echoes, reverberations and resonances. The sound is 'coloured' or 'shaped' – muffled, amplified, and so forth – by the environment it traverses (see pp. 19–23). The importance given to the acoustics of concert halls and the lighting of art galleries attests to this ecological dimension of perception. However, it is important to emphasise that 'in the absence of sight, no *new* sense develops to compensate for the loss. Through necessity, use and practice, the remaining senses become more acute and their potential is more adequately realised' (Hill 1985: 104).

Most of the examples considered in the present section are taken from the reflections of John Hull (1990). He is a university lecturer with a history of eye problems, registered blind in 1980 and who became totally blind by 1983. His account is particularly revealing about the transition from a visual to an auditory and tactile world; from the residuals of a visual memory to a totally non-visual geographical and social experience.

Acoustic space is quite different from visual space. 'Sound places one within a world' (Hull 1990: 62). Sight, by contrast, sets a world, as an object or series of objects, in front of the eyes. The visual world is somehow separate and composed as a scene, whilst the sentient participates in the auditory world and it unfolds over time and continually changes. We can close our eyes to the world, but not our ears – even the sleeper is awakened by a sound. Ihde (1976) argues that we are always at the edge of visual space and set at a distance by sight. By contrast, we are always at the centre of auditory space, listening out with the ear. This is well illustrated by the following passage from Hull (1990: 62).

On Holy Saturday I sat in Cannon Hill Park while the children were playing. I heard the footsteps of passers-by, many different kinds of footsteps. There was the flip-flop of sandals and the sharper, more delicate sound of high-heeled shoes. There were groups of people walking together with different strides creating a sort of pattern, being overtaken now by one, firm, long stride, or by the rapid pad of a jogger. There were children, running along in little bursts, and stopping to get on and off squeaky tricycles or scooters. The footsteps came from both sides. They met, mingled, separated again. From the next bench, there was the rustle of a newspaper and the murmur of conversation. Further out, to the right and behind me, there was the car park. Cars were stopping and starting, arriving and departing, doors were being slammed. Far over to the left, there was the main road. I heard the steady, deep roar of the through traffic, the buses and the trucks. In front of me was the lake. It was full of wild fowl. The ducks were quacking, the geese honking, and other birds which I could not identify were calling and cranking. There was the continual flapping of wings, splashing and squabbling, as birds took off and landed on the surface, or fought over scraps of bread. There was the splash of the paddle boats, the cries of the children, and the bumping as two boats collided. Parents on the shore called out encouragement or warning. Further away, from the large expanse of the lake, there was the different sound of the rowing boats as they swished past, and beyond that was the park. People were playing football. I heard the shouting, running feet, the impact of leather upon leather as the ball was kicked. There seemed to be several groups playing different games. Here there were boys; further over in that direction there seemed to be a group of young children playing. Over this whole scene, there was the wind. The trees behind me were murmuring, the shrubs and bushes along the side of the paths rustled, leaves and scraps of paper were blown along the path. I leant back and drank it all in. It was an astonishingly varied and rich panorama of movement, music and information. It was absorbing and fascinating.

Acoustic space is dynamic, not static. It is an appearing and disappearing of sounds, of single sounds and sounds voicing together. It is a world of nothing but action. 'Every sound is a point of activity. Where nothing was happening, there was silence. That little part of the world then died, disappeared' (Hull 1990: 62). For the sighted, the visual world provides the continuous backdrop of images, both moving and apparently static. The visual world is relatively stable and available to the gaze. However, the auditory world is more intermittent. When there is silence, or only the occasional quiet sound, the auditory backdrop vanishes. Yet, at other times there is such a surplus of different sounds that it may become quite difficult to discern spatial dimensions and the world may even feel to be closing in on the sentient, like the feeling of crowding. On other occasions, there is sufficient continuity of distinct sounds to discern a geographical world, often picking sounds out which give a great sense of distance and wide open spaces. However, the blind are dependent on the voicing of that world: that which does not speak (or is not in reach of touch) does not, in effect, exist. Again, Hull (1990: 12) gives a vivid description:

> The idea of a nice day is largely visual. A nice day occurs when there is a clear blue sky ... For me, the wind has taken the place of the sun, and a nice day is a day when there is a mild breeze. This brings into life all the sounds in my environment. The leaves are rustling, bits of paper are blowing along the pavement, the walls and corners of large buildings stand out under the impact of the wind, which I feel in my hair and on my face, in my clothes. A day on which it was merely warm would, I suppose, be quite a nice day but thunder makes it more exciting, because it suddenly gives a sense of space and distance. Thunder puts a roof over my head, a very high vaulted ceiling of rumbling sound. I realise that I am in a big place, whereas before there was nothing there at all. A sighted person always has a roof overhead, in the form of the blue sky or the clouds, or the stars at night. The same is true of the blind person of the sound of wind in the trees. It creates trees; one is surrounded by trees whereas before there was nothing.

The experience of the blind reminds us how important the dialectic between ear and eye is in our own experience of space. Without the eye, the blind experience a geographical world – specifically the wider world beyond immediate haptic experience – of surprises and sudden changes. Often sounds appear quite suddenly without apparent preparation, others cease with equal suddenness. The wider geography of the blind, that of auditory experience or acoustic space, is therefore discontinuous and often unpredictable.

The blind are highly sensitive to the masking of sounds and the confusion of myriad sounds, especially in complex soundscapes such as a city street. Here the environmental structuring of sound is especially apparent. Certain sounds can drown out others or distort the acoustic space impression. In the open spaces of parks sounds may carry with some clarity, whilst in narrow back alleys even one's footsteps are heard echoing between the walls giving a clear sense of the narrowness of the space. Hill observes that a common distortion is the wind: one participant reported, 'I can't hear anything. Nothing sounds the same', another added, 'you can't tell what direction the sound might be coming from as easily' (in Hill 1985). Auditory cues which may be so clear on a quiet walk through a park or university campus in the late evening, are largely absent in the bustle of one's own home (and in fact the home, for the blind, is often a more tactile space) or the day-time city street (Hull 1990).

Therefore, though sound does not provide a continuous or reliable source of environmental information, auditory experience can give the blind a wider geography of spatial dimensions and relationships, an acoustic space, and at particular moments offer vivid evocations of place character (see the passages from Hull, quoted below). Free from the continuous flood of visual information, the blind show a clearer understanding of acoustic space. In *If You Could See What I Hear*, Sullivan and Gill clearly separate the visual from the acoustic, tactile and other sensuous experiences: 'Sight paints a picture of life, but sound, touch, taste and smell are actually life itself' (1975: 181). This is a strong argument for a study of sensuous geographies which reminds us of the relatively neglected geographies of auditory, olfactory and haptic experience. Hull (1990: 22–24) describes how the sound of falling rain creates a complete world:

> This evening, at about nine o'clock, I was getting ready to leave the house. I opened the front door, and the rain was falling. I stood for a few minutes, lost in the beauty of it. Rain has a way of bringing out the contours in everything; it throws a coloured blanket over previously invisible things; instead of an intermittent and thus fragmented world, the steadily falling rain creates continuity of acoustic experience.
>
> I hear the rain pattering on the roof above me, dripping down the walls to my left and right, splashing from the drainpipe at ground level on my left, while further over to the left there is a lighter patch as the rain falls almost inaudibly upon a large leafy shrub. On the right, it is drumming, with a deeper, steadier sound upon the lawn. I can even make out the contours of the lawn, which rises to the right in a little hill. The sound of the rain is different and shapes out the curvature for me. Still further to the right, I hear the rain sounding upon the fence which divides our property from that next door. In front, the contours of the path and the steps are marked out, right down to the garden gate. Here the rain is striking the concrete, here it is splashing into the

shallow pools which have already formed. Here and there is a light cascade as it drips from step to step. The sound on the path is quite different from the sound of the rain drumming into the lawn on the right, and this is different again from the blanketed, heavy, sodden feel of the large bush on the left. Further out, the sounds are less detailed. I can hear the rain falling on the road, and the swish of the cars that pass up and down. I can hear the rush of the water in the flooded gutter on the edge of the road. The whole scene is much more differentiated than I have been able to describe, because everywhere are little breaks in the patterns, obstructions, projections, where some slight interruption or difference of texture or of echo gives an additional detail or dimension to the scene. Over the whole thing, like light falling upon a landscape is the gentle background pattern gathered up into one continuous murmur of rain.

I think that this experience of opening the door on a rainy garden must be similar to that which a sighted person feels when opening the curtains and seeing the world outside. Usually, when I open my front door, there are various broken sounds spread across a nothingness. I know that when I take the next step I will encounter the path and that to the right my shoe will meet the lawn. As I walk down the path my head will be brushed by fronds of the overhanging shrub on the left and I will then come to the steps, the front gate, the footpath, the culvert, and the road. I know all these things are there but I know them from memory. They give no immediate evidence of their presence. I know them in the form of prediction ... The rain presents the fullness of an entire situation all at once, not merely remembered, not in anticipation, but actually and now. The rain gives a sense of perspective and of the actual relationships of one part of the world to another. If only the rain could fall inside a room, it would help me to understand where things are in that room, to give a sense of being in the room, instead of just sitting on a chair.

This is an experience of great beauty. I feel as if the world, which is veiled until I touch it, has suddenly disclosed itself to me. I feel that the rain is gracious, that it has granted a gift to me, the gift of the world. I am no longer isolated, preoccupied with my thoughts, concentrating upon what I must do next. Instead of having to worry about where my body will be and what it will meet, I am presented with a totality, a world which speaks to me ... As I listen to the rain, I am the image of the rain, and I am one with it.

Since the environment plays a role in determining the character of the sound perceived, the blind must learn to discriminate between the myriad sound sources and their different characteristics and to recognise the many ways in which the environment can modify these sounds. In some cases the sound may be totally changed in intensity (loudness), timbre and pitch, and perhaps easily confused with other sounds. Geographical knowledge derives from the effective interpretation of these complex sound and environment interactions. Our ability to distinguish different accents in a spoken language and to comprehend different styles of human voice, or delivery, perhaps suggests that the human ear is more remarkable than the eye!

Studies of the blind often suggest that they can discern geographical relationships with sound in a kind of 'echo-location', that is, using the environmental fashioning of sound to 'read' the structure of the environment in which that sound occurs – whether it is their own footsteps or voice, or some other sound source. Hull (1990) describes a particularly subtle form of this, sometimes known as 'facial vision', which is based on the close attentiveness to tiny echoes generated by a slow moving object (the walking blind person) against its surroundings. Here attention is not so much on voices or things sounding, but on their presence as a kind of pressure on the surface of the body. This 'facial vision' gives a strong sense of presence, almost physical in its intimacy. Yet, 'one is not aware of listening. One is simply aware of becoming aware. The sense of pressure is upon the skin of the face, rather than upon or within the ears' (Hull 1990: 21). Facial vision is a very strong sense when conditions and attention are aligned. It can be described as a kind of touch or acute hearing (see Chapter 4). It is a kind of early warning system: 'I must stop when I sense something, but not sensing something does not mean that I can go ahead' (Hull 1990: 19).

The more usual perception of various kinds of echo in the environment – often set up by the footsteps of the blind person on a resonating surface – permits the blind person to make broad distinctions between open spaces and enclosed ones. However, this auditory geography is far less reassuring than haptic experience. Auditory experience, within the context of a wider multisensual encounter with the world can, however, give important geographical information, of location and spatial arrangement and a rich evocation of the distinctive character of places at different times of the day and in different seasons. In our own visually dominated geographies, we often forget the range and depth of auditory experience, as illustrated in the passage from Hull (1990: 146–147). He was invited to a country church wedding. The visual delight of the building and its setting was specially chosen by the bride and groom. Being blind, Hull was excluded from this chosen delight but nevertheless had his own distinctive experience of the place.

In an unfamiliar place ... I am not preoccupied by the thought that there are things that I cannot see. My attention and emotions are occupied by what actually presses in on me ... it was the bells. I could have stood there listening to those bells for a long time. The air was full of the vibrations. My head seemed to be ringing. The ground seemed to be trembling; the very air was heavy and springy with vibrations. I tried to count how many different patterns there were ringing and, without success, to work out how many bells must be in the tower. I thought that I must really become more expert in this lovely thing. I tried to describe the qualities of the sound to myself, mentally comparing it with other bells I had recently heard. Again and again, the descending peels chimed out, over the babble of conversation, cutting up the cool autumnal air, weighing everything with a strange, solemn expectancy. I was flooded with joy, and repeated again and again in my heart, 'yes, I hear you, dear bells, I hear you' ...

I do not know whether the sighted people even noticed the bells. At best they could have only been an extra item of atmosphere, adding to the autumnal leaves and the Norman tower as the bridal party gathered in their beautiful clothes. To me, the very air I was breathing was bell-like.

Hull (1990: 125) asks the pertinent question: what corresponds visually to the difference between sound and silence? It is not the difference between seeing and shutting one's eyes because you can always open your eyes and see again. With a perceptiveness derived from his blindness, Hull argues that with the eyes 'it is within your power to grasp again the object of sense, but when there is silence, the ear has no power to grasp the sound again' (1990: 125). We close our eyes but the object is still there, but silence means the 'object' has gone, or the event has passed. Is the difference between sound and silence like that between seeing and blindness?

This is clearly incorrect. Blindness is to sight as deafness is to sound. Blindness is an internal state. One knows that the external world is still there to be seen. One has merely lost the faculty of seeing it. In the case of silence, however, the external world, the world of sound, is not there any longer. It has gone into silence.

(Hull 1990: 125)

However, to experience silence is to wait. It is not to stop listening but to continue listening, but to hear nothing. In total darkness sight is useless. The blind have no use for eyes. In total silence, however, hearing is not useless, it is just that at the moment nothing can be heard. This is the attentiveness of the auditory system, ready to warn us of change in the environment.

There is no exact visual parallel for the distinction between sound and silence.

One never possesses a sound, one never has it within one's power the way that one possesses sight. The evil eye has power over the world, but nobody every heard of an evil ear ... The ability to close one's eyes represents the power one has over things that are seen, the power to exclude. Hearing, however, is always receptive, whether to sound or to silence. You can look away, but you cannot listen away.

(Hull 1990: 126)

One might not totally agree with this evaluation but it has much in it. Nevertheless, in different cultures and times people have attempted power with sound – the voice of God for the Jew, the exact imitations of wildlife and environmental sounds made by the Saami.

ORAL CULTURES AND AUDITORY WORLDS

Schafer has made the profound observation that 'defining space *by* sound is very different from dominating space *with* sound' (1985: 95). Dominating space with sound, such as through the use of excessive amplification of a single source of sound(s), and or through the introduction of certain monotonous sounds, quite literally destroys auditory geography since it reduces or submerges the pre-existing soundscape under the blanket of the dominant sound.

Human societies have dominated space with sound to differing degrees at different times in history. This can be both intentional, as in the church bell marking out its parish or the call to prayer from the minaret of the mosque, but at other times this domination can be more accidental, a consequence of new economic and social practices. Industrialisation and urbanisation in the West was particularly associated with a new imperialism in soundscapes (Schafer 1985). Today, one only has to try to record the sound of bird song in the central park of a city to quickly realise how much so-called background noise dominates the soundscape. Ordinarily we almost ignore this sound, but when trying to record the bird song it becomes an intrusive noise. This is chiefly the dominating sound of traffic, but also within buildings we have the continuous hum of air conditioners and heat-exchangers.

Amplification of sound not only fills space, it obliterates it. This amplification can come about by the sheer number of a given sound source, such as traffic, or by deliberate attempts to make the sound stand out, as with in-store advertisement messages or at the open-air rock concert. Sounds with a monotonous or repetitive character can have a similar effect, and all the more so when loud. In the natural world such monotonous and loud sounds do exist, such as the crashing waters of Niagara Falls, but these natural sounds are far more complex and transparent than many human generated sounds, such as those of machines and motors. The sound of the crashing falls fascinates in a way that the air conditioning motors, for instance, appear not to.

Spaces defined *by* sound are hi-fi soundscapes, rich in detail and offering a clear sense of a world around us (Schafer 1977). These are generally the more natural landscapes of wild places, perhaps also the farmed countryside prior to the mechanisation of ploughing and harvesting, and are the soundscapes experienced by aboriginal peoples in their more isolated homelands today and in Europe in the past. Dominating space *with* sound produces lo-fi soundscapes, low in detail and often appearing cacophonous or just noisy (Schafer 1977). These soundscapes are all too common in Western cities, dominated by machine sounds and sounds with repetitive and harsh timbres, and often by loud sounds or sounds which blanket out other sounds. In such lo-fi soundscapes, hearing offers little assistance, or joy, in geographical orientation in space and relationship to places.

In that vague time and place which we now so easily romanticise, before the emergence of urban-industrial societies, the era of aboriginal cultures, it is widely believed that auditory sensitivity was far richer than today and soundscapes were somehow richer. (Here the term 'aboriginal' is used to refer to pre-industrial societies, both the hunter-gatherer societies, such as former Inuit and aboriginal cultures, and basic herding and agricultural societies, such as those of the Saami and medieval Europe.) Something of this original auditory geography can be reconstructed if one considers the traditions and attitudes of aboriginal peoples recorded in history and the traditions of those surviving today. The role of sound in these cultures – in art, religion and day-to-day life – suggests a geographical experience less dominated by the visual, more multisensual and often more acutely tuned to the peculiar properties of sound in the environment, that is the auditory geography.

Aboriginal cultures were highly adapted to their environment, tuned to the rhythms of nature and knowledgeable about its intricate details. As their various creation myths amply demonstrate, they felt a close kinship to the earth. For people in modern Western cultures, the individual is apart from the environment and the earth is divided up into parcels of land which constitute possessions, owned individually or by organisations. For the aboriginals, the individual and community were inseparably linked to the earth – often to particular parts of it – and in a certain sense it 'possessed' them; it was a gift not a possession, of which they were stewards not owners. For many of these cultures, from Australian Aborigines and New Zealand Maori to the Arctic Inuit (Eskimos) and the Saami (Lapps), sound played a key part in their environmental sensitivity (e.g. Chatwin 1987; Beach 1988; Peterson 1972). As gatherers, hunters and herders, they were most able imitators of the animate and inanimate sounds around them. Their auditory geography had both listening and voice; that is, it was inclusive of great attentiveness to the sounds of the environment and an effective ability to communicate with that environment (animate and inanimate) by oral mimicry (Rundstrom 1990; Carpenter 1973).

For the Inuit, 'the binding power of the oral tradition is so strong as to make the eye subservient to the ear. They define space more by sound than sight' (Carpenter 1973: 33). The visual space of the wide, arctic plains – especially in the winter darkness and ice – is almost featureless and lacking in perspective. When asked to draw a map of their world by early travellers to the frozen North, after some puzzlement at the request, the Inuit would draw remarkably accurate maps of quite considerable areas of territory, indicating a detailed knowledge of the shorelines (Carpenter 1973; Blakemore 1981). Various explanations have been offered for their remarkable accuracy. Rundstrom (1990) attributes it to their extraordinary ability to mime or imitate aspects of the environment. 'Mimicry was institutionalised not merely as an artistic ability; it was a practical tool in everyday life …' (1990: 160). But for the Inuit, a 'map' was a temporary sketch in the earth and not a mark on paper kept to indicate land possession or control. Carpenter (1973) even suggests that such mapping for them was quite alien since their world was most clear to them in its auditory rather than visual form. This was an auditory world of events, processes and actions, not the visual world of places, patterns and objects. The wind was perhaps more important than the vista, offering environmental information from its noise, force and direction, and from its olfactory content as well. The long periods of darkness in the tundra winter and the snow and ice expanses where sky and land and sea merge make visual sensitivity less useful, especially when the individual is hidden well into his or her parka to keep out of the cold and biting wind. Instead, the other senses take on a greater importance; including the hearing of distant and invisible sound sources – the water against a shoreline hidden by fog, a subtle change in the sound of ice over which the sledge is moving, the tone of the wind as it howls. It is a land not of objects – that is the world of sight – but one of events and relationships – that is the all-round alertness of the ears.

According to Carpenter (1973), the Aivilik Eskimo did not conceptualise space and time separately but saw a situation as a dynamic whole. Their concept of space was not one of static enclosure, such as a room with sides or boundaries, but was open, unbounded and dynamic, with direction and relationship key properties. Carpenter was frequently astounded by the Inuit ability to follow a trail across an apparently featureless tundra waste, even at night, and was impressed by their distant hearing. Whilst he could not see or hear an approaching plane or sleigh, his Inuit guides were already pointing it out to him well in advance of its 'appearance'. With only rudimentary knowledge of formal mechanics and limited experience of machines in general, early visitors to Inuit communities were also impressed by their ingenuity and effectiveness in mending broken machines. They did not see machines as parts fitted together but readily appreciated a machine as a dynamic functioning whole. All these abilities suggest more of an auditory than visual sensitivity, one which is about flows and continually changing relationships, rather than about objects or parts and compositions or views.

Rundstrom (1990) notes the importance that extremely accurate mimicry played in Inuit life. Mimicry was the basis of their effective hunting on the open, exposed tundra landscape where there is no cover for a hunter to hide from his prey. According to Inuit tradition, in the distant past humans and animals spoke the same language and good hunters still know the sounds that seals, walrus, polar bears and caribou will respond to and frequently engage in 'conversation' as they approach (Rundstrom 1990: 163). (Similar beliefs are held by other aboriginal groups – see Chatwin (1987) on the Australian Aboriginals.) Inuit traditional hunting, without modern rifles and using traditional tools, involved great patience, attention to details of the landscape, intense listening and, on approaching the prey, a subtle communication between human and animal. This involved both gestural mimicry of the animal's characteristic movements and oral communication using the animal's vocabulary. This mimicry extended throughout Inuit cultural life to include art and religious belief (Carpenter 1973) and was as much a propitiation and appeasement of the souls of the respected animal as it was a hunting decoy (Lopez 1986).

The song is also important to many aboriginal cultures and often holds and maintains its tradition. For the Saami, there is the distinctive animal-like sounds of the *joik* (Beach 1988). The joik is quite literally an auditory geography – it is an oral description of the earth and things in it. It is a distinctive style of singing which early Christian explorers described in derogatory terms as sounding like the barking of dogs and the howling of the devil. The ancient joiks are indeed not unlike the barking of dogs, or wolves, or seals, or the sounds of Arctic birds, the wind and the wide expanse of the Arctic landscape. Imitation of the music of their landscape is at the root of the Saami joik. It is a relationship to their environment, both immediately emotional and immediately practical. Like many aboriginal peoples, the Saami are excellent mimics and lovers of song. The Saami shaman, the religious leaders in the pagan period, used the joik to 'communicate' with the agents of the natural world. The joik expressed – and continues to express – the intimate unity of the Saami with their land. To an outsider, the joik is a song but to the Saami it is much more.

The Saami Johan Turi has called the joik singing 'a way to remember' (Beach 1988). It is a memory of their culture and history, and a respect for the landscape which has sustained the Saami. It is sung with a very compressed voice, with tense vocal chords and a very narrow throat (Edstrom 1990). It is generally unaccompanied with limited lyrics of single words or short phrases interspersed with vowel sounds. Joiks are sung both as solos and with groups of people. They express a close bond between people and between people and their land. Joik is a spontaneous form of singing, and whilst there are traditional joiks of ancient origin, many are improvised or elaborated anew each time. The joik is more process than product. Imitation is at the root of the joik and in evoking in the singing tone and rhythm the characteristics of person, animal or place, the Saami singer feels himself close to the object, he can 'remember' it. A bear joik cannot be mistaken for a reindeer joik. The joik may have a few Saami words but often it is totally lacking in words. The joik has deep roots in the shamanic past and singing a joik can be a very emotional experience for Saami.

There is an ancient tradition of relatives composing a personal joik for a young boy or girl, and for individuals and communities to develop specific joiks at key times in their lives. An individual and a community may, therefore, be defined by a collection of joiks and even a specific way of singing. Consequently, there are wide regional variations in joik styles, especially between southern and northern Saamiland (Lappland) and the joik continues to evolve. It forms an important part of the revival of Saami culture and language in Scandinavia (see Edstrom 1990).

The Saami and Inuit define space *by* sound and not with sound. Listening and mimicry lie at the root of this intimate relationship with the environment. It is a kind of harmonic auditory geography, in contrast to the cacophonous experience of Western cultures, where ever louder sounds compete for attention and noise pollution is increasingly an issue (Tempest 1985). The Inuit and Saami concepts of sound seem more authentic to the nature of auditory experience than our own. For such peoples the world is less one of objects already given and displayed to the eye, but more an unfolding world of activities and relationships. The Inuit artist doesn't impose form but seeks to reveal it – somewhat like a phenomenologist allowing things to be in themselves (Husserl 1983) – that is, let them sing. 'He transfigures and clarifies, and thus sanctifies' (Carpenter 1973: 33). When carving a piece of ivory, the artist does not impose his choice of subject to shape, rather he turns the piece over and over until he finds its suggestion, and then carves away the excess to reveal the hidden form within. It was once quite common for an Inuit to throw his carving away once completed. It was not an object for contemplation but an activity, a process or relationship with his world. Like the Saami joik, Inuit carving was a spontaneous creation, a kind of remembering, and a intimate relationship to the world. This kind of 'world view' or sense is more akin to an auditory than a visually oriented experience.

This essentially auditory concept of reality is difficult for us to appreciate in Western cultures. For instance, in language we make clear distinction between nouns and verbs, grounding our concept of the world on the logic of the eye which identifies objects and interactions between them. In contrast, the Inuit and Saami languages make little distinction between 'nouns' and 'verbs', rather they are agglutinated languages which are expressive of moments of becoming and activity. In Inuit, all words are forms of the verb 'to be', which in itself is lacking – 'all words proclaim themselves their own existence'. 'Eskimo isn't a nominal language; it doesn't name things which already exist, but brings things/action (nouns/verbs) into being as it goes along' (Carpenter 1973: 38). This is illustrated by the tradition of naming babies at birth. As the woman enters labour, the oldest woman of the community (who remembers all the names of those who have died) recites the names of the deceased and the name that is called as the baby emerges is the name it is given. For the Inuit, the baby answered to its name.

These aboriginal peoples did not have a concept of space as an empty void which might be filled with objects, but it was a space already filled, an acoustic space. 'The essential feature of sound is not its location but that it be, that it fill space' (Carpenter 1973: 35). This contrasts to the Western concept of space. This is graphically illustrated in the tale of the Inuit hunter who when told that an oil can was empty, struck a match and peered inside and bore the scars for life! 'Empty' space made no sense to them. All space has life or living forces of some kind, therefore all space has sound, or voices, which define it.

The infiltration of modern Western culture into these aboriginal societies has both undermined and transformed these oral traditions. Today, aboriginal art is sold to tourists, whether found in Australia or the Arctic tundra, and sometimes it is even manufactured in quantity in factories (sometimes far away from the region concerned) specifically dedicated to the tourist trade. Songs, such as the joiks, have become pop records and symbols of the revival of a community identity. However, though the auditory distinctiveness of these communities has survived, their oral tradition has been replaced by a literary one, and today the television aerial increasingly dominates the skyline of aboriginal settlements. The visual is increasingly as, if not more, important and increasingly impacts on the whole nature of their thinking.

Recently, it has been observed that Western culture itself may be returning to the auditory dimension and rediscovering (or redefining) its oral tradition. Since the Renaissance, Western culture and science, seems – at least superficially – to be predominantly visual (see Chapter 7), yet in the twentieth century and especially in the last few decades, auditory means of communication and oral culture have gained increased prominence. The television is both a visual and auditory medium, informing us about the world around us and influencing our perceptions of the environment, people and things in it. Furthermore, the telephone has become such an important part of our community that letter writing seems, some argue, to be in decline. McLuhan (1962) was one of the first to recognise this revival of oral culture in the West, but it is important to remember that this is quite a different kind of auditory geography to that experienced by aboriginal cultures or by our forefathers in pre-industrial Europe. In terms of Schafer's distinction (1985) between defining space by sound and with sound, it does seem that we continue to employ sound to dominate space, to give it particular characteristics or to assert our control over it. The oral culture of the West is one of musak, loudspeaker announcements, ghetto-blasters, machine warning signals, and the ubiquitous dictatorship of the telephone bell. This is very different from the receptive, patient, mimicry of the aboriginal auditory geography. Maybe we are rediscovering our auditory geographies in the West, but this is not a revival of something long since lost, but rather yet another redefinition of the role of the sense of hearing (and the voice) in geographical and social experience.

> Auditory space has no favoured focus. It is a sphere without fixed boundaries, space made by the thing itself [soundfield] not space containing a thing. It is not pictorial space, boxed-in, but dynamic, always in flux, creating its own dimensions moment by moment. It has no fixed boundaries, it is indifferent to background. The eye focuses, pin-points, abstracts, locating each object in physical space against a background; the ear, however, favours sounds from any direction.
>
> (Carpenter 1973: 35)

7

VISUAL GEOGRAPHIES

INTRODUCTION: VISUALISM

The term visual geography is used to highlight the distinctive contribution of the visual perceptual system to geographical experience, that is, our location and orientation in space, spatial relationships and the characterisation of places. It could be argued that the term 'optic geography' might be more accurate since this would specifically identify the eye as key factor. However, since the visual is such a dominant part of our making sense of world around us, it is useful to use the more general term 'visual' which can allow us to explore not only the geographical experience which is generated immediately by the act of seeing with the eyes but also the visual tools (such as the map) and metaphors (such as landscape) which we regularly employ in encountering the environment.

Vision is an important part of everyday experience of the environment and in a very real sense one can argue that geography is a kind of making visible of the world, its features and processes, both literally and metaphorically, as a contribution to understanding our place in the world. However, despite the regular use of maps and photographs, visual descriptions and diagrams, to present geographical information, we often forget how much of that information is more than what is seen. Vision is very much a taken-for-granted sense, and its true nature and limits are almost hidden by its visibility. This is the paradox of visual geographies: on the one hand it is one of the most familiar dimensions of our experience of the environment and on the other hand we know so little of how it gives us a sense of space, people, things and place in conjunction with the rest of our sensory faculties.

'Geography is to such an extent a visual discipline that, unique among the social sciences, sight is almost a prerequisite for its pursuit' (Pocock 1981a: 385). *Sensuous Geographies* has already indicated that this widely accepted assumption is something of an exaggeration, but it is quite true that geographers have tended to neglect the non-visual dimension of human experience of the environment (Porteous 1982) and until relatively recently paid little critical attention to the visualism of the discipline (e.g. Cosgrove 1984, 1985). Much of geographical understanding is grounded in 'visualising', that is using and making maps and satellite images, observing behavioural patterns and processes of physical change, representing complex relationships in diagrams and other visual models. Perhaps one of the most important parts of fieldwork is *seeing* places first hand, and in the past the ability to draw a sketch was regarded as a vital skill. We often forget that this seeing is also touching, smelling and hearing the environment which we explicitly or implicitly compare to our own and previous experience. Training the eye, or 'an eye for the country', that is, its physical structure or human organisation, continues to be a valuable part of geographical education. From Cornish (1928, 1935) to contemporary humanistic geographers (Rees 1980; Tuan 1979a), it has been recognised that the geographical epistemology is predominantly a visual one.

Cosgrove (1984) links the 'visual bias' in geography to the use of the concept 'landscape' and argues the detailed descriptions of earlier geographers – such as Vidal de la Blache, Carl Sauer and Jean Bruhnes – had a visual foundation. This is to mean that not only do geographers favour seeing and visual tools over other techniques of exploration and recording, but the way they conceptualise and think about problems and issues is conditioned by this visualism. Even taking a far more extended view of the history of the discipline (e.g. Buttimer 1993) to the earliest writings that might be called 'geography', the description of the earth (*geo-*, earth; *-graphe*, drawing) has been an attempt to visualise, or make visible, both the features and then forces, patterns and processes, operating in the world around us (Holt-Jensen 1981).

Jean Bruhnes's distinction between the artist and the geographer illustrates the 'visual foundations' of geography:

> the perspective of the geographer is not that of an individual observer located at a particular point on the ground ... The landscape of the geographer is very different from the painter, poet or novelist. By means of a survey, sampling, or a detailed inventory, he achieves the comprehensive but synthetic perspective of the helicopter pilot or balloonist armed with maps, photographs, and a pair of binoculars.
>
> (quoted by Mikesell 1968: 578)

Here, we not only have an eye-first geography, but an eye linked to synthetic thinking and the geographer employing visual tools. The whole statement is written within the framework of visual metaphors – landscape and perspective.

This visualism, or tendency to reduce all sensuous experience to visual terms, is evident both in geography as a discipline and in contemporary culture (Eco 1986). It is, however, a relatively modern phenomenon and closer examination suggests that such visualism is not so all pervading as might initially be supposed. The visual might be a dominant sense mode but it is not independent of the others and is not necessarily the most important

116

in many situations. Sensuous geographies vary between individuals due to their physical condition (age and health, relative disabilities), education and cultural context. Any understanding of visual geography must be grounded both in the physiology of vision and cultural definitions of its use. Visual geographies are ways of making sense of the world in which the eyes play an important but not exclusive part.

SPACE AND VISION

Joseph Addison wrote in 1712, 'our sight is the most perfect and most delightful of our senses' (*Spectator* 1712, quoted by Relph 1982). It is commonly claimed that humans are first and foremost visually oriented, though closer inspection reveals such claims to be exaggerated. Nevertheless, vision is particularly important in geographical experience. Sight gives us a synthetic view of the environment as a whole, as a view or scene, and allows us to differentiate objects in terms of their colour and texture, shape and form, relative size and arrangement in space. However, sight is so much part of our everyday experience, that we tend to take it for granted and do not fully appreciate its distinctive role, nor its subtle inter-relationship with the other senses.

In spite of the richness of detail, range and variety of visual information we are able to acquire, the apparent coherence of a synthetic view of the environment inclusive of stationary and mobile objects, and the rapidity with which we are able to recognise and evaluate features from their visual appearance – it is important to emphasise that last phrase – sight is concerned with appearances. On its own it gives us access only to surfaces. The rest of the perceptions we associate with vision are either supplementary percep- tions from the other senses – haptic, olfactory, auditory – or from our previous experience and memories, or speculations. The eye is a wonderful mechanism which gives us a vast array of information but we must not overestimate its geographical role. Visual illusions are some of the com- monest. Vision is a deductive sense and it is a dependent sense. It deduces – 'works out' – the nature of objects and spaces from surface information. It depends upon the information of the other senses and the memory to assist this interpretation of visual images or surfaces. The eye is also reliant on light. In the dark our visual geography is far more impoverished.

Sight is dependent on the pattern of light and the arrangement of surfaces with respect to that light (Cornish 1935; Gibson 1974). Sight offers a geography of surfaces. Sight offers visual representations: these are appear- ances of phenomena in light but not, strictly speaking, the phenomena themselves in their fullness and depth. The possibility of illusion is always present. Seeing is a creative interpretation of appearances, a translation of what appears as patterns of illuminated surfaces into what is represented, that is people and things. Visual (and auditory) geographies are relatively abstract and metaphorical and, in this sense, unlike the geographies of touch and even smell which are more direct and interactive. The visual world is 'out there' beyond the eye, suspended in some hypothetical distant space quite unlike the immediate tangibility of haptic and olfactory geographies. Nevertheless, the possibilities opened up by sight are considerable and the visual dimension forms an important part of our geographical experience.

Sight, like the other senses, mediates person–environment encounters. It enables us to distinguish individual objects, their arrangement in space, and their key surface characteristics, including their size and form relative to our own bodies and their position and movement relative to other objects in space and ourselves. Sight also offers us a whole world or scene, giving both a sense of a stable background geography and a foreground of greater detail and potential or actual movement. Just as we might hear particular sounds or recognise a melody, so we see particular objects or identify a scene. However, this geography of appearances is set at a distance as an 'out there' and we can quite easily position ourselves as a detached observer of such a scene. Sight is therefore both abstract and synthetic. It not only records light sensations but, in conjunction with the brain, composes them. Visual perception involves both the immediate physiology of the eye and the activity of the brain (expectation, memory, analysis, synthesis). Giving us a 'map of surfaces', sight allows us to orient ourselves in the environment and, through its combined properties of synthesis and detachment, gives an overall view of a world. Interestingly, the idea of a mental image, or map, of places and environments (e.g. Gould and White 1974) and the legibility of environments (Lynch 1960) is generally conceived in visual terms. This reminds us that tendencies towards synthetic and detached seeing are strongly conditioned by our cultural tradition, that is, we in part must learn how to see in this way and it may not be the only way seeing can operate. The 'codes' of visual cues are interpreted in socially agreed ways (Berger 1972, 1980). Visual geography is therefore not totally explained by physiology, it is also culturally defined, and the visual experience both a personal and a shared experience.

It is useful to remember the sensuous matrix (Figure 7.1) outlined in Part 1. We use our sense of sight both to receive information and, either directly or indirectly through some other sense organ, present ourselves as visible to others. There is, therefore, a perception and a presence, a gathering of a world and a participating in that world. Sight, like the other senses, gives us access to a geography and through our own visibility we have a geographical presence. We can further distinguish between passive and active modes of attention in each sense mode, though in practice this appears to be a continuum of degrees of attentiveness/non-attentiveness in environmental encounter (Seamon 1979).

Although vision is so important to the adult, it is not available to the pre-born child in the womb and even young children seem to be more tactile and

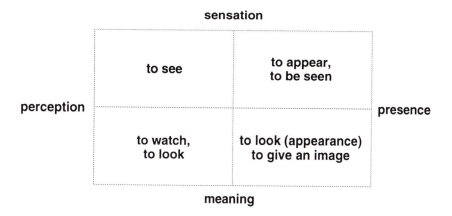

Figure 7.1 The visual sensuous matrix

olfactory than visual in their orientation. It is also evident that in aboriginal cultures vision is not so dominant as it appears to be in the West. For the traditional Inuit, 'the binding power of the oral tradition is so strong as to make the eye subservient to the ear. They define space more by sound than sight' (Carpenter 1973: 33). Their conception of the world is more auditory than visual. This is why, perhaps, when asked by early European explorers to draw 'maps' to show their routes and pathways the Inuit appeared quite puzzled. They had not a map drawing tradition nor really a visual conception of wider spaces. The tundra Arctic which constituted their home is often wide and featureless, and dark for much of the year, so that a purely visual concept of space would be a poor substitute for auditory, tactile and olfactory information about that landscape – the direction and force of a wind, the odours in that wind from sea, ice, animals and vegetation. In Western culture sight can seem so dominant at times, that we only really become aware of its all-pervading quality when considering such other cultures and when studying the experience of the blind (Hill 1985). Sight is perhaps the one sense in contemporary Western cultures that people most fear losing. To be blind is to be lost in space, it is assumed.

The basic physiology of the eye is illustrated in Figure 7.2. The eye consists of both a receptive surface (the retina) which transmits (or translates) light information, photons, to the brain as nerve impulses and the associated muscles which rotate the eye and orientate it towards various light surfaces. The retina surface consists of a mass of light sensitive cells. These are most compact near the centre and generate the fullest information from this point. It is common to distinguish three areas on the retina surface (Gibson 1974) which are responsible for three kinds of vision.

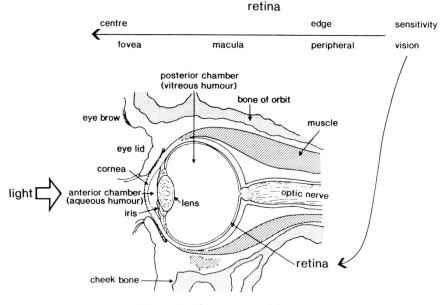

Figure 7.2 The structure of the eye
Source: developed from Gibson (1974) and Long (1992)

1   The *fovea* is associated with sharp vision of features ¹⁄₉₆ to ¼ inch at a distance of 12 inches from the eye. This is the kind of vision which enables one to thread the eye of a needle and is an invaluable part of sight.
2   The *macula* surrounds the fovea and covers a visual angle of 3° in the vertical plane and 12-15° in the horizontal. This vision is quite clear but not as sharp as foveal vision since the light-sensitive cells are not as closely packed. It is particularly important in colour sensitivity.
3   The *peripheral vision* is associated with the edge of the retina where the light-sensitive cells are least dense. This peripheral vision provides a basic detection of movement 'out of the corner of the eye' and is therefore important in redirecting the attention of the eye as a whole.

In the healthy eye, these three types of visual image are experienced simultaneously and blend with one another to give an overall three-dimensional, textured and dynamic visual world. The healthy eye is able to discern distances in space (depth), spatial relationships (arrangement of objects in space) and distinguish fine details about the specific character of surfaces (or objects) and much more valuable geographical information.

Gibson (1968, 1974) proposes an ecological theory of perception and specifically illustrated this with reference to vision (see Part 1). His early research arose out of the need to improve the spatial perception of pilots

120

flying over targets and coming into land with their planes. Through this work, Gibson came to realise that we cannot explain visual perception (or for that matter any other form of sensuous experience) purely in terms of the structure and function of a sense organ, such as the eye, nor by some 'black box' model of interpretative calculations made by the brain. Gibson recognised a third vital variable: the structure of the environment itself, which mediates the sensuous information reaching the sense organs. His explanation of visual perception is essentially geographical since it recognises the role of context (the structure of the environment) in perception and attends to the problem of spatial perception and orientation.

Gibson argues that vision is not merely the reception of light sensations by the retina which then passes them by nerve impulses to the brain to be processed into visual images, but rather vision is a process which involves the eyes receiving an 'optical array' from which it discerns the colours and textures of surfaces, the form and arrangement of objects, and distances, depths and movements of things in space. The optical array or bunches of optical arrays are reflected light, that is radiant light which has been structured by the surfaces from which it is reflected. If one looks at the sun, or a light bulb, one sees only whiteness and can gain little visual information about the source or the environment. This direct light is the radiant light. The eyes are a specialised organ for 'reading' the ambient or reflected light. In other words, in an illuminated environment we are able to see objects and their surface characteristics because light is reflected from them. Darker surfaces reflect less than light ones, different textures reflect differently, and overall variation in reflectivity across our view gives a 'picture' of an environment of different objects distributed in space. The ambient light can come from a number of sources as it is reflected from surfaces at different angles to the illumination (or other various sources of illumination) and, of course, as we move our eyes and head, and even our whole bodies through the environment the range of structured visual information – optical arrays – increases and the clarity of our vision improves. This is why a picture can sometimes be quite obscure yet it is only in special circumstances, such as poor lighting conditions, that our eyes are confused in direct visual experience of the environment. The photograph represents the visual experience by reducing it to a two-dimensional image of colour or tone variations. This ecological perspective therefore conceives of visual perception as an active and exploratory process, and visual space as a continuous but variable surface illuminated by light from an external source (such as the sun). Position relative to the light source and reflectivity of surfaces enables the form and depth of surfaces to be discerned. If the term 'space' is used to represent the structure of the environment, that is, the sum of people, things and the distances between them, we can summarise Gibson's theory as effectively arguing that we perceive spaces because space structures the ambient light reaching our eyes, and so vision is spatial.

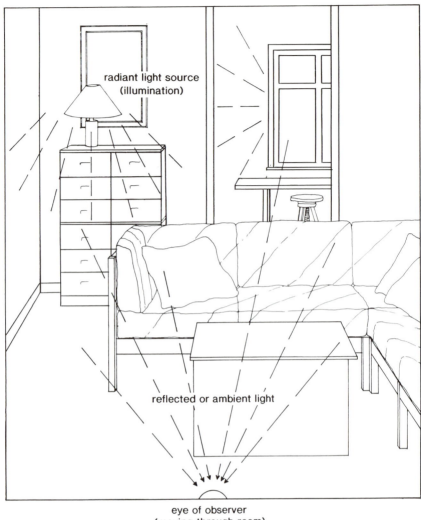

Figure 7.3 The visual system or vision as spatial

However, the relationship between vision and space is not just a physical or ecological question, it is also a cultural and even political one. Aesthetic theorists and feminists in particular have drawn attention to the social history of visual perception, that is, the ways in which we use our eyes and interpret our visual experience is culturally defined and is learnt by a socialisation into the culture. Feminists have highlighted the genesis and consequences of particular cultural practices of sight (Rose 1986; Pollock 1988). Specifically, they have argued that in Western culture vision has been defined in terms of masculine hegemony, whereby the male eye defines the visual style and the female, and specifically the female body, is positioned as viewed object or spectacle. Power lies in the all-seeing eye, so that such a constitution of the visual implies a hierarchical dichotomy between the dominant male and the subservient female. This therefore suggests that visual geographies, and geography as a discipline also, are characterised by a masculine gaze and, furthermore, throughout much of the last 300 years or so that gaze has also been upper or middle class and from the white male. This expression of power, this practice of visual perception, is most clearly seen in art but is also evident in the wider visual experience of society as a whole, including concepts of the human body and geographical space (Rose 1986).

The eye is male, the object observed is female. 'Investment in the look is not as privileged in women as in men. More than other senses, the eye objectifies and masters. It sets at a distance, and maintains a distance' (Irigaray 1978: 50). The very idea that to see is to have control over the thing seen might be considered a masculine (and white male) concept. Furthermore, emphasis on the visual – the appearance, the image, the surface – displaces the material and participatory for the cognitive and detached. 'In our culture the predominance of the look over the smell, taste, touch and hearing has brought about an impoverishment of bodily relations. The moment the look dominates, the body loses its materiality' (Irigaray 1978: 50). In other words, a dominance of the visual reduces the body to a surface and marginalises the sensuality of the body. This same argument can also be extended to spaces in general, that is, the environment becomes first and foremost visually experienced and other sensuous geographies are marginalised. Historically, therefore, the dominance of masculine vision has led to assumptions such as that which is not visible is less important, and that which is not visually pleasing to the male is inferior. Feminists contrast this to what is believed to be a more rounded sensuous experience characteristic of the feminine sensitivity. Here, the visual and auditory are more closely integrated with a stronger tactile and olfactory composed experience of the world, one which values nurturing and cooperation. Whilst this feminist analysis can seem to exaggerate and stereotype visual practices, there is strong evidence to support many of these arguments, especially when our own experience is compared to that of non-Western cultures where the visual is defined quite differently. Feminists therefore give us an important reminder of the cultural definition of sensuous experience. In an analysis which disregards gender issues, Appleton (1975) does recognise a hierarchical relationship between the viewer and what is seen. In the prospect–refuge theory of landscape perception this is explained by reference to our evolutionary past as hunters. Feminists focus on more recent documented history to explain gender differentiation between the viewer (voyeur) constituted as masculine and dominant and the object viewed (a body or a landscape) which is defined as feminine and subservient.

Therefore, it seems that the eye, as defined by the Western cultural tradition, sets at a distance what is viewed, detaches the observer from what is observed and implies an hierarchical relationship of power giving the viewer – who chooses to look – a kind of control or privilege over what is viewed. Furthermore, this visual geography is presented as 'objective', as a verifiable truth. However, appearances deceive and illusions can be both accidental and quite deliberate. This is the duplicity of vision. (Duplicity here is used to refer to the contradictions within our concept of vision and differs somewhat from Daniel's (1989) socio-political concept of duplicity.) Whilst touch is perhaps the most truthful sense – within its range of accuracy – vision is perhaps the most easily fooled – as the continued fascination of visual illusions well demonstrates. Sight both sets at a distance, offering a view to the observer which is detached and claims objectivity, and yet sight also composes that view or scene or perspective by being a selective interpretation of appearances or visual representations. Sight chiefly gives information about surfaces in the environment and changes in those surfaces, but not on its own direct and full information about objects and relationships between objects (things and people) in space. This composition of the view is socially and personally determined by education and cultural practice, that is, it is essentially 'subjective'.

TIME, SPACE AND SIGHT

When looking at photographs or remembering a scene or the look of someone, it is easy to fall into the assumption that sight is about static images, about objects and their arrangement into compositions or pictures. This snapshot concept of visual experience does not encapsulate the everyday experience of sight. Sight is amenable to being remembered as images, looks, views and scenes, just as we might remember an experience of sounds as a melody or tune, but it is experienced as a dynamic encounter with the environment, a complex mixture of apparently static elements (such as objects at rest in specific locations) and moving elements (see Gibson 1968, 1974).

Visual images are multidimensional and continually changing. We perceive the motion of objects relative to one another, and specifically objects (people, things or substances) moving against a relatively static background. The distinction between foreground and background, which is so basic in visual geography, is a distinction between movement and apparent stability. The stability of the background is sometimes more apparent than real, since things in the far distance may be moving but we do not perceive them as such. From an aeroplane flying thousands of feet up above the ground at great speed, the ground can appear to be a stable image, yet looking out of a train travelling rapidly along its tracks, but considerably slower than the plane, we are very conscious of our movement through the landscape and a tendency for some closer images to blur. Individual perception of space with sight is, therefore, one of relative movement, or stability against motion, which is partly realised through the distance between foreground and background, and between our own motion and that of the objects observed.

The blur, or visually indistinct image, is as much part of visual geography as the clear image. Blurring may be due to the rapid motion of an object which is so fast that we can't catch a stable image of it, or because it is beyond our field of vision, as commonly experienced by the short-sighted when not wearing spectacles or contact lenses.

Considering visual geographies as dynamic also introduces the notion of time. Visual 'objects' subsist in both space and time, they have location relative to other 'objects' and have duration relative to other 'objects' too. The film is perhaps a closer analogy to visual geographies than the photograph. Whilst observing the film, we are aware of things moving and time passing, of missing some details which have 'passed by' and of lingering on the relative constancy or slower changes of others. Vision is not presented with a picture of a totality to view at leisure, to explore methodically like a work of art, but rather visual experiences flow past us, we catch glimpses of this and that, identify and linger on this or that, and so build up a collection of images and changes in our minds, that is, we compose a view. In remembering experience, we tend to reduce the flow of visual experience to specific images or scenes, that is to moments, to snapshots.

In considering examples of ways in which we use sight in geographical experience and visual tools and concepts to present geographical information, such as the map, or to simulate the experience of places, such as film, we continually need to remind ourselves about the dynamic nature of visual experience, that is, of its temporal nature. Vision is time-specific, in the sense of when an object is illuminated in a particular way at a certain moment, and temporal in the sense that visual images persist over time and give continuity to geographical experience. The blind have a quite different experience of time, in part, because of their lack of sight (Hull 1990). Without the continuous backdrop of visual experience (when our eyes are open and the illumination is adequate), geographical experience can become more fragmentary. To give continuity, haptic experience must be active and exploratory. Olfactory and auditory experiences are always immanent or potential until the moment of occurrence and duration arrives. The geographies of olfaction and auditory experience can be fragmentary, and to the blind who are denied the anticipation of vision, can be sudden encounters or assaults on the body.

Vision gives both a geographical and temporal continuity. It gives a time–space unity to experience of the environment. On awaking from sleep, especially sudden waking, we can be quite disoriented in part because of the break of vision. Rapidly circling around on the roundabout, visual images pass by all too quickly for us to fix them in a moment in time (or sequence) and space (or location) and so the image is blurred and we become dizzy. When the time–space coherence of vision is suspended, as in sleep, or broken, as in rapid motion, we quite literally lose our place in the world, spatially and temporally.

PERSPECTIVES AND SYMBOLS

Every culture develops key concepts and tools which effectively represent the shared understanding of particular features of their experience. Of particular interest to visual geographies are 'landscape' and 'map'. Both are concepts – ways of organising geographical experience – and tools – techniques for recording information and expressing a certain level of understanding. Landscape and map both emerge as concepts, tools or visual strategies in forms in which contemporary users would recognise them during the Renaissance. Researchers have traced a history of both forms much further back, in the case of the map, to pre-history, and in the case of landscape, to an early form of land ownership in post-Roman times. Yet, such proto-forms are perhaps not really what they seem. For this reason, we concentrate here on the modern development of these forms and consider them as important indicators of ways of seeing and thinking visually in European (and American) culture since the Renaissance.

Landscapes and maps, therefore, are culturally specific visual strategies, ones which have evolved and changed over time. Landscape came to refer to a special kind of perspective, a way of seeing, which first developed in art and has continued to be important in photography and film today. The map appeared from more varied antecedents, including verbal and written itineraries as well as attempts to chart or map journeys and territories. A key characteristic of the 'coming together' of the map, or cartography, was the development of techniques of visual symbolisation. Despite the differences between landscapes and maps as visual strategies, there are many parallels in their development, and they have in practice complemented one another and reinforced a specifically modern European view of the world, a visual geography which seeks objectivity and distance from what is viewed, and composes, organises and expresses a control over what is made visible. In this sense, they are structures which mediate our experience. They present kinds of simulated geographies, composed of specific visual perspectives, conventions and symbols. They both reduce the complex multisensual experience to features represented visually *and* organise, compose or synthesise these into 'scene' or meaningful whole.

The perspective of landscape

The term 'landscape' is sometimes used as if it was synonymous with the term 'geography' and there is a long history of geography as the study landscapes, both physical and human. More recently, there has been a tendency to invent terms which refer to specific types of landscape – bodyscape, inscape, smellscape (Porteous 1985, 1986a, 1990), soundscape (Schafer 1977), flatscape (Norberg-Schulz 1969: 268), townscape (Cullen 1961), cloudscape and seascape. In everyday speech we commonly use the term 'landscape' as a substitute for 'countryside', 'environment', 'place', and even refer to urban 'landscapes'. This wide use of the term, however, tends to lead to dilution of its meaning and above all encourages us to forget that its use in all these contexts is metaphorical. The term 'landscape' is a complex concept which has undergone significant changes in meaning and connotation, both in its more general use and within geographical application. Landscape illustrates the importance of the visual in person–environment encounter, and the way in which cultural practices define the visual. Landscape is a kind of microcosm of the history of visual geographies in our culture.

First and foremost it is important to emphasise that landscape is a socio-historical construct. Today the concept appears highly amorphous since it has been used in a variety of different ways. The use of this term in geography illustrates the way in which the origins and myriad associations of the term 'landscape' have been neglected as people have used this apparently innocent term. More recently, a number of geographers have begun to explore the history of visual geographies and to question what is meant by the term 'landscape'. There is a large literature on the landscape concept in geography (Meinig 1979; Relph 1981; Cosgrove 1985; Daniels 1989, etc.), here only a small proportion of this literature is considered.

Below are a number of definitions of and reflections on the term 'landscape'.

Landscape is a kind of backcloth to the whole stage of human activity.
(Appleton 1975: 2)

'Landscape', as the term has been used since the seventeenth century, is a construct of the mind as well as a physical and measurable entity.
(Tuan 1979b: 6)

A landscape is a cultural image, a pictorial way of representing, structuring or symbolising surroundings ...
(Cosgrove 1984: 1)

When we consider landscape, we are almost always concerned with a visual construct.
(Porteous 1990: 4)

Landscape is not merely an aesthetic background to life, rather it is a

127

setting that both expresses and conditions cultural attitudes and activities, and significant modifications to landscapes are not possible without major changes in social attitudes ... Landscapes are therefore always imbued with meanings that come from how and why we know them.

(Relph 1976: 122)

A working country is hardly ever a landscape. The very idea of landscape implies separation and observation.

(Williams 1973: 120)

... when collective labour and the struggle with nature had ceased to be the only arena for man's encounter with nature and the world – then nature itself ceased to be a living participant in the events of life. Then nature became, by and large, a 'setting for action', its backdrop; it was turned into landscape, it was fragmented into metaphors and comparisons serving to sublimate the individual and private affairs and adventures not connected in any real or intrinsic way to nature itself.

(Bakhtin 1986a: 217)

Landscape is a social and cultural product, a way of seeing projected onto the land and having its own techniques and compositional forms; a restrictive way of seeing that diminishes alternative modes of experiencing our relations with nature.

(Cosgrove 1984: 269)

Sometimes a landscape seems to be less a setting for the life of its inhabitants than a curtain behind which their struggles, achievements and accidents take place ...

(Berger 1972: 13, 15)

Landscape came to mean a prospect seen from a specific stand point.

(Tuan 1974: 133)

Landscape is everything I see and sense when I am out of doors ... landscape is the necessary context and background both of my daily affairs and more exotic circumstances of my life.

(Relph 1981: 26)

Landscape is an attractive, important and ambiguous term.

(Meinig 1979: 3)

These quotations suggest a wide range of different yet tantalisingly related meanings and associations attached to the term 'landscape':

1   an area, space, or scene;
2   a relationship between people and environment;
3   a visual representation;
4   it implies an entity or whole;
5   a way of seeing; a technique;
6   a physical backdrop, yet . . .
7   its form reflects social attitudes;
8   it implies the separation of an observer and an object, and
9   an illusion, a curtain or cover-up of the reality of everyday struggles.

Each of these formulations of the idea reflect the complex history of the concept/tool and in use of the term are often combined in many different ways. A simple definition of 'landscape' is difficult to formulate without excluding at least one or more of these connotations.

Houston (1982) identifies three related terms: *landskrift* – a unit of area that is a traditional entity, such as the lands of a tribe or feudal lord; *landscape* – associated with the Dutch painting tradition, a unit of visual perception, a view or scene generally depicted with much realism; and *landschaft* – a German term which includes both these senses and has much influenced geographical usage and is perhaps a source of much of the confusion and ambiguity in the term today. Tuan (1974: 133) traces the word landscape back to the Dutch *landshap* which referred to commonplaces such as 'a collection of farms and fenced fields'. It was this term, he argues, that was later transferred to England in the sixteenth century and applied to art. Here it became associated with concepts such as the picturesque. Cosgrove (1984: 9) traces a different history of the term beginning with the artistic use of 'landscape' from the early fifteenth century to the late nineteenth centuries, to denote an artistic and literary representation of the visible world – scenery which is viewed or contemplated by a single spectator from a particular point in space. This landscape was a particular way of seeing the world, detached, abstract and synthetic, and implying a relationship of power, the observer is dominant over the observed. Cosgrove contrasts this to a twentieth-century geographical and environmental use of the term to refer to the integration of natural and human phenomena, which can be empirically verified and analysed by the methods of scientific enquiry, over a delimited portion of the earth's surface. The aesthetic and scientific usage both aim at a kind of realism. The scientific is a literal realism, a verifiable one. The artistic is a creative realism, a simulation and not verifiable since it deliberately excludes certain features and adds others to heighten the effect of realism.

In the history of the term landscape, it is possible to discern three incarnations. First, there is the *material* term, perhaps the original use, which is hidden in the etymology of the term *land/scape* and related *landshap* and *landskrift*. This refers to a unit of area, such as feudal lands or tribal territories. The '-scape' can be traced to the suffix '-ship', as in friendship and

129

kinship, and refers to a state of being (Relph 1981: 26). This therefore suggests a root meaning of the term landscape as a material unit of area which is worked or held by a group or an individual – who are in an established relationship to that piece of land or territory.

Second, more widely recognised is the *aesthetic* term, but this is often poorly understood. This refers to a technique in art commonly associated with the landscape painting which emerged in Flanders and Italy from the early fifteenth century and was a widely accepted tradition in Western art into the late nineteenth century and continuing in photography today (Stechow 1968; Herman 1973; Rosenthal 1982; LAC 1986; Berger 1972, 1980). Here, landscape is a visual composition of a scene depicting a place or area, either in imagination or reality. However, even the most realistic landscape paintings modify the scene to improve its visual impact and almost never reproduce exactly an actual visually experienced space. The basis of this visual representation is the technique of linear perspective which enables artists to reduce a three-dimensional visual experience to a two-dimensional image, a realistic view from a specific point (Cosgrove 1984, 1985). The form of the landscape, or its structural aspect, is focused upon here, that is linear perspective. However, a further part of the visual presentation is the use of colour and texture to describe an emotional response to the scene depicted. The more affective dimension is especially a feature of nineteenth-century Romantic painting and is also reflected in the choice of types of landscape. Landscape painting in earlier centuries tended to depict countryside and village domestic scenes, whilst painting of the later period included wild and barren mountains and moors, and even bleak and stark representations of the urban and industrial environments, often in the rain. The artistic use of the landscape concept implies a certain kind of relationship to the world, the eye of the detached observer. It is also an early form of simulated geographical experience (see Chapter 9).

Third, the *geographical* term emerges, more ambiguously, initially as a term used in physical and regional geography at the start of the twentieth century and more recently adopted and modified by humanistic geographers. The initial use tended to emphasise an identifiable physical unit, associated with geology and geomorphology, patterns and processes, sometimes described as 'natural regions' (Herbertson 1905). Later, landscape came to represent in geography a synthesis of physical and human elements in the constitution of a distinct region, sometimes described as functional or economic (Johnston 1983a: 42–49). More recently, drawing more directly on art, humanistic (behavioural, perception and cultural) geographers have associated the term with the aesthetic qualities of the environment, or scenery, and with the emotional experience of places and regions (Relph 1981; Tuan 1979b).

Therefore, the term landscape offers a visual geography which composes the environmental experience into discrete portions of land, area or scene – real or imagined – and establishes a certain kind of relationship to it –

material or aesthetic. Everyday use is undoubtedly dominated by the aesthetic use of the term, with its emphasis on visual representation. This is important since the reduction and recomposition necessary to depict a portion of the earth's surface, or our experience of a part of the environment, to a two-dimensional image and for the viewer to contemplate that image (painting or photograph) and assume it is 'realistic', or at least an adequate representation, demands a particular style of seeing and attitude to the world. This is the technique of linear perspective and the attitude of a detached observer. This is quite specific to our cultural tradition and contrasts markedly with other cultures, such as the aboriginal Inuit, for instance, who did not have a concept of linear perspective but a more multidimensional concept which favoured no one perspective (see Carpenter 1973). The evolution of the term landscape implies an increasing abstraction in the relationship between person and world in Western culture and the establishment of a kind of power relationship, not only between viewer (with privileged viewpoint and power to compose the view) and object (that seen), but also between artist or expert (e.g. architect, geographer) and other observers (or consumers) of the painting, photograph or sight-seeing tour. We perhaps forget that it is because of the established tradition of linear perspective and landscape that we can experience photographs and television pictures, in particular, as 'realistic' representations of people and places. The relationship implied becomes less and less material and more and more representational and metaphorical. And as with all such widely used metaphorical structures, we begin to forget that the concept and tool we call 'landscape' mediates between us and our direct encounter with the environment, it translates our experience into a particular kind of visual geography. (This argument is closely related to Baudrillard's (1983a) orders of simulacra which identifies three stages in the evolution of representation: counterfeit, production and simulation. This will be followed up in Part 3.)

Cosgrove (1984, 1985) argues that 'landscape' is a way of seeing which grew out of Renaissance humanism's desire for realism in art. The technique of linear perspective emerged as a way to represent the world as a visual impression and parallels developments in map-making which were at the same time attempting to generate more realistic records of the world in visual form. (In Gibson's (1974) ecological theory of visual perception, linear perspective is but one of thirteen varieties of perspective: perspectives of position (texture, size, *linear*), perspectives of parallax (binocular, motion), perspectives independent of the position or motion of the observer (*aerial*, blur, relative upward location, shift of texture or linear spacing, shift of rate of motion, completeness or continuity of outline, transitions between light and shade). Aerial might refer to the perspective found in the map (see also Hall 1969: 179–182).) Whereas in map making the view from above and practices of symbolisation developed, in painting the view from a particular point on the ground and an attempt to draw and represent the details of three-dimensional space in two-dimensional pictures developed. Although distinct, the two developments – landscape and map – represent a similar progression towards visual abstraction: one pictorial and the other symbolic.

Linear perspective effectively defines landscape as a kind of reflection. In his *Della Pittura* of 1435, Leon Battista Alberti demonstrated 'a technique for constructing a visual triangle which allowed the painter to determine the shape and measurement of a gridded square placed on the ground when viewed along a horizontal axis and to reproduce in pictorial form its appearance to the eye' (in Cosgrove 1985: 48). This gave the realist illusion of a three-dimensional space on a two-dimensional surface. Such a strategy makes form and position in space relative rather than absolute, that is, the forms of what we see as objects in space and geometrical figures vary with the angle and distance of vision. The eye is regarded as the centre of the visual world, the sole mediator of appearances. This technique expresses a control over space. It is a realist representation of space within which the individual is both the central reference and controlling factor. We turn our eyes and the world changes; there is a fresh perspective. 'Visual space is rendered the property of the individual, detached observer, from whose divine location it is a dependent, appropriated object' (Cosgrove 1985: 49).

Therefore, landscapes are about objects and volumes but maps are about

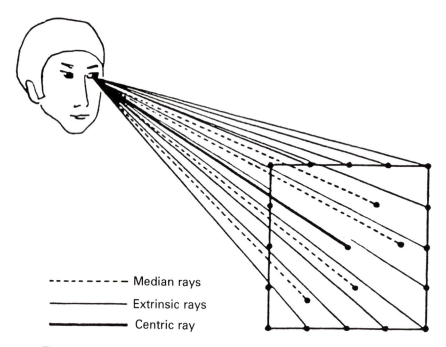

- - - - - - - - - Median rays

——————— Extrinsic rays

━━━━━━━ Centric ray

Figure 7.4 Linear perspective. The visual triangle as described by Alberti
Source: Edgerton (1975) in Cosgrove (1985)

points and distances. The landscape painting makes form and position in space relative rather than absolute. The map – through symbolisation and techniques of projection – similarly depicts relative relationships proportional (in scale) to the original ones found in our visual encounter with the environment. Although maps depict what is actually visible, they also can 'visualise' what is not visible in everyday experience, and through the selectivity of the map-maker certain elements are shown and given relative importance whilst others are not. Unlike the painting, the map is a more abstract visual composition, a view from a vertical rather than horizontal plain, usually drawn at a constant scale across its surface and so privileging no one point of view. Like the painting, the map is a selective composition, but whilst the painting depicts a discrete place, the map covers a continuous space which is always potentially part of a much larger space (and/or map). However, ordinarily, we use maps with little concern to the specific nature of the visual geography that they represent and the way they have evolved into the abstract forms we use today.

Visual abstraction and the map

The history of the 'map' is illustrative of the history of ways of seeing and of the evolution of the geographical perspective. For many people, the map is geography. The map represents its description of the earth, its synthesis. However, the study of maps and the techniques of making effective maps are the concern of cartography. For geography, the map is one of many tools used to present geographical knowledge, but to some extent it is for geography what the manuscript is for music.

Today, we clearly see that the map is quite distinct from the picture and maps are so commonly used that we often forget that it is a highly specialised form of representation (Wood 1993). Not all cultures have maps, nor do they use map-like artefacts in the way that we do. The origins of the map are somewhat obscure and any attempt to chart the evolution of map-making is inevitably retrospective and describes certain artefacts in ways in which their creators perhaps would not have conceived them (e.g. Harvey 1980; Crone 1966; Bagrow 1985). Nevertheless, there is a rich and interesting literature on the history of map-making (in addition see Tooley 1978; Thrower 1972; Baldcock 1966; Robinson and Pechenik 1976; Cuff and Mattson 1982). The geographical literature on maps is quite varied, ranging from maps as tools (Board 1967; Blaut 1971; Guelke 1976; Harley 1983, 1989/1992), to particular kinds of map – especially so-called 'mental maps' (Gould and White 1974) and maps in education (Borchett 1987).

Maps are characterised by a number of features:

1   The map presupposes *a certain view of the world*, a specific style of visual geography, one which takes a kind of bird's eye view.

2   The map is a *scale drawing* not an exact reproduction. On early maps the scale varied across the map, but on modern maps it is usually constant across the whole surface.
3   The map is also *symbolic*, that is the myriad features of the environment are depicted, or represented, by agreed symbols – figures, lines, shadings.

Whilst a few lines and points drawn in the sand to give the rudimentary directions to a traveller may represent a proto-map, the satellite image with its complex colour simulation of distant surfaces represents the ultra-map, or even a kind of post-map.

'The purpose of a map is to express graphically the relations of points and features on the earth's surface to each other. These are determined by distance and relation' (Crone 1966: xi). This is a technical definition of the map – one among many – but even a brief survey of histories of map making and everyday uses of the term 'map' suggest a far more complex concept. Harvey (1980) notes the lack of a word for 'map' in ancient European languages and the ambiguity in such words which have been applied to map-like artefacts – pictura, figura, effigies, maps (or mappa), carte, itineraries and charts. Spanish, English and Polish derive their label from the Latin *mappa* meaning cloth – by way of the 'mappa mundi', a cloth painted with a representation of the world. Most other European languages derive their label from the late Latin *carta* which meant a sort of formal document. The resultant ambiguity persists today in words such as the French *carte* and the Italian *carta* which can mean other things beside a map. Harvey (1980) notes similar ambiguities in non-European languages. The Arabic *nuq shah* which is the basis of the map word in most Indian languages can mean picture, general description or even official report as well. All these ambiguities reflect the range of uses and associations attached to maps and map-making, such as the world views represented for the Christian church and other religions; charts and maps for trading and voyages of discovery; surveys and scientific treatise; and the demands of military and political administrators. In other words, the map is a cultural product. Today the map is considered an accurate visual representation, drawn to scale and using conventional symbols and signs. It represents features and relationships in the environment, it is used to locate things or to find a route, and it is used to establish claims to space. Each of these characteristics is a historically specific manifestation of the 'map'. 'Maps' have not always been understood or used in such ways. Furthermore, many applications of the term 'map' to various map-like artefacts of the past and of different cultures, especially pre-1500, inevitably represent the reinterpretive standpoint of hindsight and our Eurocentric perspective.

The map as we might recognise it today, seems to begin to emerge most clearly around about the same time as the concept of landscape was being

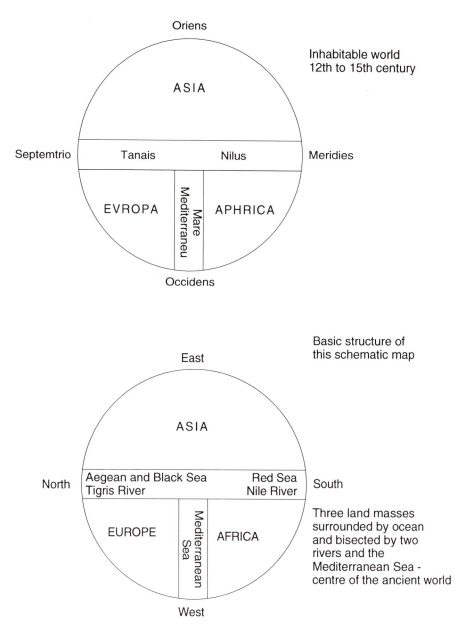

Figure 7.5.1 T-O map

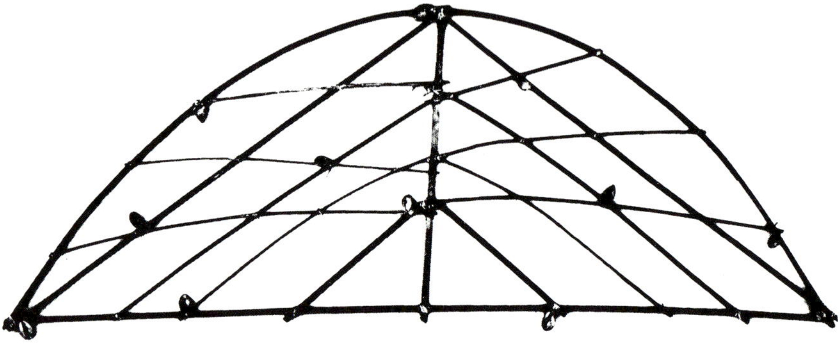

Figure 7.5.2 Stick chart, Marshall Islands (shells marking islands are joined to sticks that represent currents and tides)
Source: Harvey 1980: 29

developed, that is post-1500. As already noted, there are many similarities between these two forms of visual geography, but the key distinction seems to be that landscape emerged as an artistic technique whilst map-making emerged as a practical tool for merchants and government officials. Maps can be considered at various scales, and in particular the history of world maps can be distinguished from that of more local maps. Here, we concentrate largely on the latter. Figure 7.5 shows examples of various kinds of map. In modern times maps have been used to represent ideas, to indicate ownership or possession, to record physical features and resources, and for way-finding or recording routes. Here we concentrate on examples of maps which describe the environment and record locations and their inter-relationship, but even this limited selection offers a variety of forms: picture-like maps, charts and itineraries, maps of Ordnance Survey kinds and schematic maps and diagrams like that of the London Underground.

The map is a visual tool but it is not a picture or photograph. Of course, maps both today and in the past have been enjoyed for their visual beauty, as aesthetic objects as well as practical tools. The painting or photograph generally aims to reproduce actual shape, colour and texture of features and is not so abstract as the map (with the exception of abstract art) and like the map reduces people and places to a two-dimensional image. The map tends to cover a much larger area than the picture, and replaces actual features with a set of more or less arbitrary signs and symbols by which they are displayed far more clearly and far more uniformly than in a picture or photograph (Harvey 1980). The map selectively represents the environment and may visually represent both actually visible features, ones which would be hidden from a viewer on the ground and even non-visual features. The map can also overlay other demarcations, such as names for features and the lines of political boundaries. A map, like other 'texts', can also be read as a source of

136

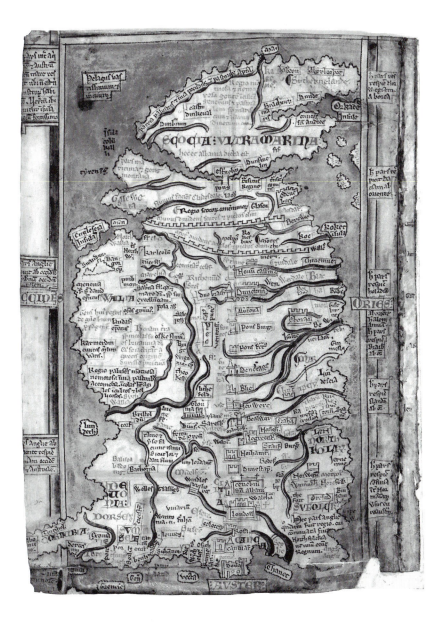

Figure 7.5.3 Matthew Paris's Map of Britain, mid-thirteenth century
Source: British Library Reference RR7494

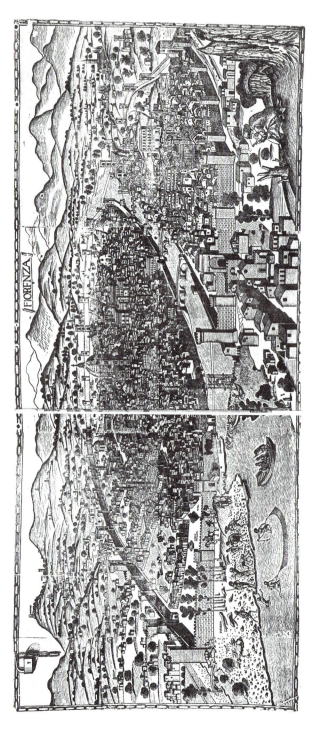

Figure 7.5.4 'Map with Chain', Florence 1482
Source: Harvey 1980: 68–69

138

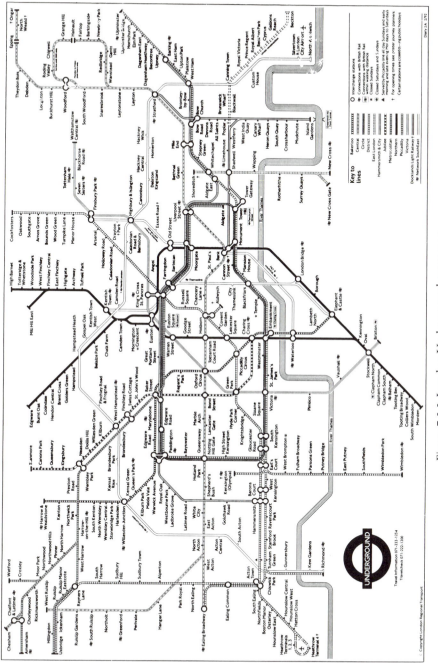

Figure 7.5.5 London Underground route map

Source: London Transport Museum. RT Registered User No. 93/E/596

139

information about its creator and the cultural context of its creation. The map both represents the environment and presents the cartographer at one and the same time. Maps are therefore records of historical geographies and records of past perceptions, that is, sensuous geographies (and visual ones).

The map therefore symbolises and adds texts to the visual geography. The map can include features such as contours – lines drawn to represent points of equal height – as well as symbols to represent particular buildings and land uses, and can also indicate relative importance by the colour, size or even design of symbols. The map therefore presents visually both features and relationships in space and ideas or claims about or on that space. It represents human priorities and knowledge.

The map is a kind of snapshot, it is a given space at a particular moment in time. It is a static visual geography. The map depicts patterns of natural and human features in the landscape, or environment. Processes or motion may be represented on maps – such as trade-winds, river flow directions. However, the map cannot capture change like a film – except in a series of separate maps showing the sequence of changes over a period of time.

It is not surprising therefore that maps come in a variety of styles: symbolic, pictorial, schematic and various combinations of these. This variety is a record of the role of the visual in geographical experience and the evolution of visual representation. It is tempting to presume an evolution of visual representation from the highly symbolic representation of early religiously inspired maps (such as the T-O maps of early Christian Europe, see James 1972: 70) through pictorial maps (such as the bird's eye view town plan maps of Italy, 1500–1650) to representational maps (such as a modern Ordnance Survey map). However, this oversimplifies the complex inter-action of sensuous experience and cultural practices and is highly reductive. Simple grand chronologies are ill-advised in tracing the development of map styles, though in particular periods certain types of map can be related developmentally, that is chronologically. Thus one finds examples of symbolic, pictorial and schematic mapping styles in most periods. (The typology symbolic, pictorial and schematic is not dissimilar to Baudrillard's (1983a) orders of simulacra – counterfeit, production and simulation – considered in Part 3.)

Briefly, several important changes in the visual geography of the map can be recorded which may or may not be causally related (see Harley 1983; Bertin 1979; Baldcock 1966). From aboriginal times, it seems, humans have made use of *symbolic representation* side-by-side with *pictorial representation*, that is drawings of crosses and circles to represent particular things, places and events, and pictures or figures for people, animals, mountains and trees (Harvey 1980). The modern development of the map (since about 1500) seems to be a transition from pictorial elements (actual picture-representations of town layouts) through picture symbols (such as a tower for a town) to abstract symbols (such as crosses for churches, different sizes of dot for different sizes of settlement). This progression suggests an increasingly abstract visual representation but it also is associated with increased detail and sophistication in the map. The comparison is most readily seen when comparing the bird's eye view of Florence (1482) with an Ordnance Survey map (see Figure 7.5). A characteristic of the modern map – Ordnance Survey-style or schematic like that of the London Underground – is the use of predominantly abstract visual symbols (shapes, colours, etc.) established by social convention. (Computer mapping is not considered here. See Aalders 1980; Taylor 1980; Monmonier 1981; Schofer 1975; Hsu and Porter 1971.)

The second feature of the modern map is the *perspective* adopted. Harvey (1980: 48) writes: the pictorial maps are distinguished as maps, not just pictures, 'because they represent landscape features as if seen from one view point unattainable in reality'. This is not the linear perspective of landscape painting, though its emergence is contemporaneous in Italy and related technically. The viewpoint is unattainable in reality and so demands an imaginative transposition. In earlier maps, such as the 'Map-with-the-Chain' (Florence, 1480s), this is a view from a hill above the town which does not actually exist, but later, as in the Venice map of 1500, it is literally a bird's eye view, the perspective is from above. This view from above becomes the dominant perspective in modern Western maps. Its emergence seems to be closely associated with the move from pictorial representation to more abstract visual symbolisation (Harvey 1980). Early forms of this view tend to have the scale changing across the map, reflecting a single, central viewpoint, but gradually, as accuracy of maps becomes more and more important, a constancy of scale is established across the whole map surface, effectively giving no one vertical view any privilege over the others. The bird's eye view, and its transition to an overall vertical view with no privileged viewpoint, also echoes the growing mastery of human society over the natural world and the rise of powerful elites and absolute rulers in Western societies. The map in this guise suggests a detachment from the world and a power or control over the world in the hands of the map owner (or reader) which previously was the preserve of God. This is perhaps to speculate, but nevertheless the bird's eye view is a different style of use of the eye to everyday visual experience down on the ground in amongst the houses and streets, farms and fields of day-to-day life.

Third, and related it seems to both the increased *visual abstraction*, detachment and power over the world, in the interests of realism and accuracy, maps in the modern period have increasingly been drawn to *uniform* scale. In earlier times, the scale on maps was highly variable across the surface, with features close to the observer being generally larger than those farther away, and features deemed of greater significance generally drawn larger than those of minor importance. We can still see this today in sketch maps drawn by both children and adults. The increasing use of maps for navigation or way-finding and the legal record of territory, with the rise of trade and empires perhaps lies behind these technical developments in scale drawing. The purpose of scale drawing is primarily to preserve distances between locations, so that the visual representation accurately reflects the actual geographical relationships. Scale was applied on two levels: this preservation of relative distances and in the systematic grading of symbols to reflect relative importance of particular features, such as the size of settlements and the importance of roads. Scaling not only miniaturises, it also abstracts. In scale representation, one usually conceives of distances as they absolutely are, constantly, across the map, unlike actual experience where accuracy of distance perception declines with distance from the present point. However, what is perhaps most important is that the map with a constant scale removes the privileged location of the observer. The painting and the bird's eye view still preserve the actuality or necessity of an observer at a given or privileged view-point. The constant scale map, such as a sheet from the Ordnance Survey, has no privileged observation position: the map makes sense whether looked at in one part or another, whether viewed one way round or another. Obviously, we prefer to view maps with the North at the top and the text written on the map oriented similarly so we can read it, but this does not deny that, apart from the names and other writing on the map, it can be read from any angle. Further, as the O.S. series illustrates, the sheet division becomes arbitrary, areas of map can be selected to suit particular needs, such as tourist maps which consist of parts of several standard O.S. sheets so giving the tourist a more economic choice of map for his/her needs. The London Underground map also illustrates, in more abstract form, the same lack of a privileged viewpoint. One can buy a whole network map, or view portions of the routes in each tube train.

The modern map, therefore, presents a visual geography of abstract symbols, detached perspective and uniform scale which constitutes a culturally unique form of visual representation and, in the effective use of maps, visual experience. Landscapes and maps each play a key role in our experience of the environment and our thinking about the world, they each mediate our visual experience and constitute an important dimension of our visual geography.

Part III

SENSE AND REALITY

8

SENSUOUS GEOGRAPHY AND TRANSFORMATION

INTRODUCTION: TRANSFORMATIONS OF SENSE(S)

In Chapter 1, 'sense' was described as having a dual aspect, containing both a reference to sensations and to meaning. In other words, sense is both biologically grounded in the physical structure of the body and its relationship to its environment (which has changed little in thousands of years) and is conditioned by the technologies (machines, buildings, etc.) and cultural practices employed by human societies in a given place and time (which have frequently changed, sometimes in as brief a period as a few decades). The 'reality' which a given culture accepts and perpetuates is closely related to the ways in which it defines sensuous experience and employs each of the senses in encountering and seeking to make sense of the environment.

The sensuous reality is determined, therefore, not merely by raw sensations or naive experience but within the context of a complex of a culture's systems of beliefs and within the confines of its technological prowess. Cultural practices and technologies of a society effectively mediate person–environment encounters and largely determine geographical understanding. In the middle of this complex the senses are both medium and message, physically and culturally defined, a structure and information. When the information of the senses seems to agree with the expectations of a given culture, that is, is compatible, it reinforces them. When the information of the senses conflicts with expectations, that is, is incompatible, the information is either not comprehended, or is ignored or dismissed as an anomaly or illusion – or contributes to a doubt which fuels new ideas and practices. It is because cultural practices change and technologies progress that this compatible/incompatible division continually changes and in so doing the senses are redefined, the style of sensuous experience is transformed. The transformation of the senses is at the same time a transformation in, or redefinition of, what constitutes 'reality'.

In the history of Western cultures, one can observe important changes in systems of belief and cultural practices, and radical transformations in technology with considerable impact on day-to-day life and ways of thinking. Therefore, it is inevitable that sensuous experience has been transformed, both with regard to how each sense is defined and employed, but also the relationship between the senses and their associated geographies – haptic, olfactory, auditory and visual. This transformation of the senses and of sense (or reality) is complex yet common and recurring themes can be discerned. There is a differentiation of the senses and each society seems to give each sense a set of key roles and interrelates the senses in certain ways, often hierarchically favouring one sense over another. Technological ingenuity permits different societies to extend the reach of certain senses, such as the telescope giving a view of distant stars and planets, and transforms the way those senses are utilised and valued. In particular, one can observe at least four related processes:

1   *Symbolisation* – a symbol is something that stands for or represents something else. The process of symbolisation is the translation of specific sensuous experiences with their distinct character into other forms, that is the remembering, recording and representing of sensuous experiences in terms of symbols. Language is a key means by which sensuous experiences are symbolised, and in particular metaphorical language.

2   *Association* – specific sensuous experience becomes correlated with particular situations and with certain emotions. Such associations change over time and from culture to culture, but can persist often long after any real or actual correlation, if it ever existed, has long since ceased.

3   *Abstraction* – to abstract is to separate from or detach; and it is also to summarise, epitomise and so reduce. In the context of sensuous experience, abstraction is a process of reduction of the range of a sense dimension which is perceived, that is a restriction of its definition to a limited number of features. Abstraction may on the one hand lay claim to contain the 'essence' or essential feature(s) of a sense but on the other it separates out and elevates certain elements of the sense but neglects others. It may also include attempts to simulate the experience, such as the synthetic odours generated by the perfume industry.

4   *Re-assignment* – each sense dimension, and even particular features of each sense, are assigned specific significance and roles within sensuous experience. These assignments change over time as one or more senses (or features of a sense) are marginalised, whilst one or more other senses (or features of a sense) gain prominence. This is also a recognition that sensuous experience is not symmetrical, that is, each sense is not assigned an equal role and the senses (and the features of each sense) are related hierarchically. This hierarchy varies from culture to culture, even from situation to situation within a culture, and such assignments undergo a continuous process of adjustment.

Each of these processes of transformation are explored here and differences

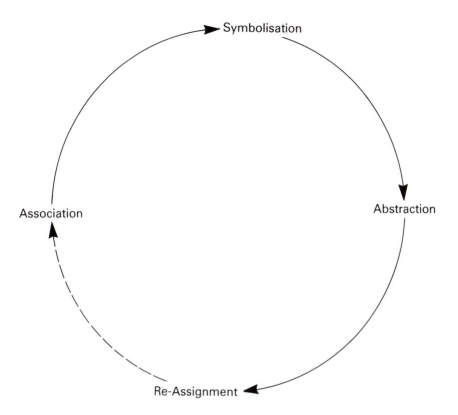

Figure 8.1 Processes of transformation

and similarities between each of the sense dimensions noted. Each of these processes of transformation could be considered as a kind of 'hyper-realisation of sense' (see Chapter 10).

Whilst it is possible to identify sequences of transformations, simple causal explanations oversimplify the complex ways in which styles of sensuous experience and concepts of reality emerge, change and, importantly, coexist and inter-relate at a given time and specific contexts. The history and geography of the transformation of the senses is therefore complex and any sketch of these transformations must be provisional. Indications of these transformations can be found directly in the explication of sensuous experience and more indirectly in the interpretation of the sensuous dimensions and priorities inscribed in place, that is the design of buildings and the organisation of space (e.g. Norberg-Schulz 1980; Tuan 1993).

REACH, CONTACT AND INTIMACY

Haptic geography is perhaps the most fundamental and basic geographical experience (see Chapter 4). It gives us a location in a world and orients us with respect to features in the immediate part of that world. This tangible and immediate world is, however, so commonplace and so continuously part of our being in a world that we ordinarily tend to take for granted much of the haptic experience, paying more of our conscious attention to visual and auditory geographies.

Williams argues that studies of tactile behaviour in different cultures show that 'the transformation of tactile experience into abstract conceptualisations would seem crucial to understanding the way some cultural conceptions are acquired by the individual in the course of cultural learning and transmission' (quoted by Montagu 1971: 240).

In Western cultures, haptic experience is often assigned a rather marginal place in the palate of sensuous experiences and its significance underestimated. It is often reduced to a reach-touch of the fingers and an extended touch of tools. In this definition, the haptic experience becomes primarily one of points and discrete surfaces or areas – that is objects – rather than contact with a continuous and varied tactile blanket. This haptic geography is one of reach and a certain distance between the sentient and the world, in contrast to the potential of touch and movement (the haptic system) to offer a world which embraces the individual within itself and gives a continuous and extremely rich geography of belonging and participation. Therefore, the assignment and abstraction of the haptic experience in Western culture suggests a kind of separation, distrust and even alienation from the physical world. The tactile experience is supplemented by and even substituted with other sensuous experience, notably seeing.

The apparent reduction of haptic experience to a more limited reach-style touch and the tendency to assign it secondary or subservient status, also seems to be related to a further tendency in Western cultures to bifurcate sensuous experience. This bifurcation is the division of the senses into two classes: so-called cerebral senses, sight and hearing, have been separated from so-called bodily senses, such as touch and taste–smell. The two classes are positioned in hierarchical dichotomy. The cerebral senses are deemed primary and superior to the latter. It is the cerebral senses that are emphasised in studies of environmental experience, information and geographical understanding. The bodily senses are not denied but are deemed secondary and inferior. In aesthetics, science and everyday life, the bodily senses tend to be given less attention and the geographical knowledge they can impart is largely ignored. Whilst it is readily recognised that young children make much use of their haptic senses to gather primary information about their environment, it is not long before the child is reminded that 'do not touch' is a recurring theme in our culture.

148

Nevertheless, touch still holds great significance in Western culture. Whilst the wider tactile awareness is somewhat neglected, touching is elevated to special significance in certain situations and has gained some quite culture-specific associations. Haptic experience is, in fact, more significant in our culture than might initially be supposed. It is very much the hidden sense – sometimes quite literally. Touch is especially associated with intimacy, trust and truth. It is used to confirm other sensuous information and to affirm contact between people and between people and their environment. Touching is closely associated with detailed evaluation of quality. Thus we feel the freshness of fruit and the quality of cloth. If we are unsure whether to believe our eyes, we will 'touch it and see', so verifying the truth like doubting Thomas.

The situations in which we are permitted to touch and the degree of tactility allowed are often quite precisely defined. These rules differ somewhat between Western cultures, from the hugging and kissing Italians to the formal handshakes of the traditional English. Yet in each case, touch implies trust and friendliness. Places where touching is permitted are generally friendly places, and in the case of North Europeans and North Americans this is more often the home or the bedroom rather than the street or public square. This restriction of tactile behaviour to intimate relationships and places, is reinforced by a whole range of other sensuous definitions of behaviour in different contexts. Therefore, in shops we are discouraged from too much handling of the goods, especially squeezing or tasting without prior permission. In a sculpture exhibition the blind seem to be the only visitors who explore the pieces with their hands, the sighted just view the exhibits like paintings. Public spaces, outside and in buildings, are often designed to give sufficient space for people to move around freely without being squashed together and when forming a queue or choosing a seat or place to stand waiting in a public space many North American and English people keep what they feel is an appropriate distance of between 1.5 and 4 feet or an empty seat between them and the next person (Walmsley 1988: 91). By contrast, the home and the regular workplace can be a much more compact and intimate space where people more readily crowd together without feeling uncomfortable (Hall 1969; Sommer 1969).

There is also a rich metaphorical use of touch defining it as relationship, contact, intimacy and verified truth. We 'keep in touch' by letter or telephone, we can become 'out of touch' with a subject, or become 'touchy' about some personal matter made public. Yet, this intimate touch is a discontinuous geography, one of occasional contacts between people and people and environment, it is supplementary rather than primary. The full potential of haptic experience to generate a rich and continuous geographical understanding is not, it seems, fully realised.

The emergence of this style of haptic geography could be described as a kind of domestication of the sense of touch and movement. The clothes cover much of the body and so restrict or muffle global touch. This emphasises the fingers as primary organs of touch and tactile exploration. The definition of particular styles of touch behaviour and its restriction to specific situations represent attempts to control touch, just as medicine is employed to control the ailments of the body. Technological aids play a key role in this domestication of touch. The invention of the pencil and the evolution of the pen, printing, typewriter and wordprocessor, not only permitted humans to communicate over greater distances and to record their thoughts, it also allowed humans to distance themselves from those thoughts by translating them into visual form – writing and diagrams.

The telephone and fax continue this transformation in the auditory and visual realm respectively, widening the geographical spread and speed of human communication (see McLuhan 1962). Such technological transformations transform haptic experience into other kinds of sensuous experience, but the reach of 'touch' is increased. Therefore, whilst haptic experience is supplemented and even substituted by visual and auditory means, its geography is extended and transformed also. Keeping in touch today is not merely visiting a friend in a neighbouring town, but it can also be a telephone conversation to Australia. It is also interesting to observe how the Modernist architecture of concrete and glass high rise structures tended to offer rather limited tactile variety and emphasised a scale much larger than the intimate geography measured by the human body. One might suppose that architecture is a sensuous science and should attend to haptic as well as visual geographies, yet Norberg-Schulz in his phenomenology of architecture notes that humans *visualise* their understanding and express their existential foothold through built forms (1980: 17).

ATROPHY AND SYNTHESIS

Lefebvre has noted that

> everyone who is wont ... to identify place, people and things by their smells is unlikely to be very susceptible to rhetoric ... the sense of smell had its glory days when animality still predominated 'culture', rationality and education – before these factors, combined with a thoroughly cleansed space, brought about the complete atrophy of smell. One cannot help feeling though that to carry an atrophied organ which still claims its due must be somehow pathogenic.
>
> (1991: 198)

Yet, olfactory experience is still an important part of contemporary Western experience.

When we explore olfactory geographies it seems that what might at first appear to be an atrophy of the sense of smell – a reduction of its significance and effective contribution to geographical experience – is in fact an 'atrophy' of a specific style of olfaction (see Chapter 5). The transformation of the senses suggests that the atrophy argument is an oversimplification. Instead, closer reflection suggests that smell has been redefined, that is reassigned, within the spectrum of sensuous experiences. We do not carry an atrophied organ but rather we use that organ in quite different ways to our forebears. Tuan (1993) identifies something of this history of smell. Here, this transformation of the sense of smell and changing olfactory geographies in Western culture will be briefly considered.

Olfactory geographies have radically changed since prehistoric times, and even since the Renaissance. In particular, as with so much of human experience, we appear to have somewhat domesticated much of the olfactory geography of day-to-day experience. Whilst natural smells – the farmyard animals, the blossom on the trees, the salt of the sea – remain in the 'outside' world, in the 'inside' world of our own bodies, our products and our buildings (and perhaps also cities) olfactory experience is increasingly controlled and organised. This organisation includes several related strategies:

1   *Cleansing* – the removal of smells, especially strong smells, by washing and other tools of dissipation.
2   *De-odorisation* – the removal of unwanted smells by the addition of masking smells (incense, deodorants, sprays).
3   *Synthesis* – the manufacture of 'bottled' smells, either as extracts from natural odour compounds or, increasingly, the generation of new, synthetic smells (perfumes, odours).
4   *Labelling* – the adding of a specific smell to a product or space, replacing any previous odour profile by a new one with desired properties and culturally agreed associations.

These closely related strategies – cleansing, deodorisation, synthesis, labelling – have tended to simplify olfactory experiences and restructure olfactory geographies, or smellscapes. This restructuring is most evident in contexts where humans have greatest control over the olfactory realm, such as the odour of our bodies or the smells of manufactured products, such as toilet cleaners and shampoos. There is a long history of attempts to control, or domesticate, body odours by using perfumes or deodorants. There is also a long history of attempts to control smellscapes within buildings. This has historically been done by burning incense, the heat spreading the fragrance of a chosen substance. The flowers brought into the house or the use of scented pomanders continue this tradition. More recently this process has been simplified by the spray can. However, the smells of the wider environment, the city streets and the countryside continue to roam free of direct manipulation.

In the outside environment, the transformation of olfactory experience is less direct and more an unplanned consequence of changes in economic and

151

social practices. In moving from farming to manufacturing, from villages to cities, from horses and carts to cars and lorries, we have replaced one smellscape with another to a great extent for much of our everyday lives. This olfactory change has not come about by design, unlike the use of a perfume or deodorant, but is more an accident. Yet it is also a consequence of our marginalisation of the olfactory geographies. Whilst we have attended to the appearance of our cities and the noise of our towns, only more recently have we begun to attend to olfactory pollution in the streets of our cities and infiltrating into our homes (Berglund and Lindvall 1979; Lindvall 1973).

Therefore, in considering the olfactory geographies in contemporary experience, it is important to distinguish between the 'outside' and 'inside' smellscapes, those which are essentially undomesticated and those domesticated. In the former, natural and complex odours predominate – especially in rural areas and wilderness areas. Changes in farm production and technology, however, have altered even these smellscapes. The odours of the poultry sheds of modern factory farms and the modern milking barns with the greater use of machines and concern for cleanliness, the odour of slurry stores and the ubiquitous oilseed rape, have each changed the olfactory geography of rural England. Urban environments have perhaps experienced the most radical olfactory changes, often regarded quite negatively. Here, the smellscape is increasingly dominated by the odour of oil-burning machines. In urban environments, strong odours of traffic and industry – notably chemical industries – have become associated with pollution and have tended to reinforce a negative attitude to smells. It is almost as if 'to smell' inevitably means 'to stink of something awful', especially in an urban context. The flower beds in cities seem to be planted more for visual effect than an olfactory experience.

The closely related strategies of cleansing, deodorisation, synthesis, and labelling can also be described in terms of the processes of transformation of sense introduced earlier. Together they suggest an abstraction of odours from their immediate context, the promotion of particular associations to certain odours and the reassignment of those odours to certain objects and situations symbolically. In a broad sense, olfactory geography is transformed from simple natural associations of smells and original sources to complex artificial or created associations of simplified smells with specific situations, emotions or ideals. Not content to accept the natural odours of products and places, or the natural odours of extracts from nature's products, we extract such odours from their natural context and association and reassign them to new roles and associations, creating new compounds of odours (perfumes) and even generating totally synthetic odour compounds (Gibbons 1986).

At the risk of over-generalisation, one can identify a process of transformation in the way in which smells are related to and defined in Western societies, especially since medieval times, and consequently changes in the olfactory geographies perceived and the evaluation of them. There would appear to be transformations from a simple naturalism to a complex symbolisation of smell resulting from a detailed differentiation of odours and the classification of these odours qualitatively, emotionally and hierarchically. These transformations increasingly separated the odour from its specific source and, importantly for a geographical understanding, from its unique location. The ability to capture or extract smells from nature's products and the development of first a trade in exotic fragrances (especially from the East to the West) and later the application of odours to human products, such as shampoos and soaps, to give them specific associations, often of archetypical geographical significance – such as the 'pine fresh' and 'spring flower' fragrances – rather than of specific locations, reinforced a general abstraction of odours and the assignment of symbolic meanings. Hence, particular odours become associated with 'the exotic', the forest or the garden, and so on.

Commodified smells – those odour compounds developed for 'olfactification' of marketed products, from soaps and detergents to fast foods – are increasingly a-geographical, lacking a specific place reference. They are

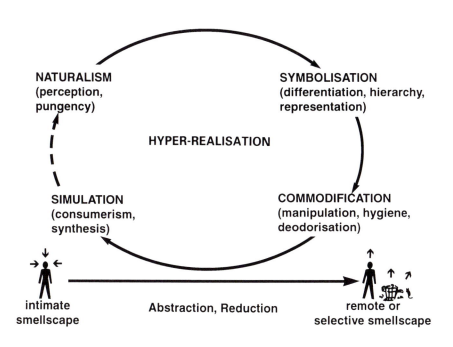

Figure 8.2 Hyper-realisation of smell

synthetic smells, either compounds of extracts from nature and/or compounds of totally synthesised odours invented by scientists. The olfactory experiences these commodified odours offer are simulations, and when employed specifically to evoke place experiences, such as in heritage experiences like York's Jorvik Viking Centre, one moves from symbolisation and representation to a realm of hyper-reality and simulation, where the odours experienced seem more real than real (see Chapter 10).

It would seem that, as in aboriginal societies, the medieval (and pre-medieval) Europeans tolerated and even enjoyed a much richer and stronger olfactory geography. Both in the countryside and the towns, the natural smells of animals and people, vegetation and farm produce characterised everyday experience. Different places would be distinguishable by their strong characteristic smells associated with patterns of agriculture and early industries. (Something of this olfactory place geography survives in the contrast between the smell of the older quarters of ancient cities, especially Arab ones, and the modern downtown quarters of Western cities.) This was a period before the obsession with personal cleanliness and the reduction of odours by much washing and before the infiltration of the smells of steam-powered engines and later oil-powered machinery. This earlier mode of olfactory experience might be described as a form of naturalism. Here natural smells predominated, specific smells were clearly identifiable to specific objects (natural referent), and there was a rich olfactory geography of different people and places. However, we have little evidence of how people actually related to odours and their role in everyday life. Comparison to the practices of aboriginal peoples perhaps give us some clues (see Chapter 5).

Whilst a simple chronology is unrealistic and oversimplified, it does seem that as human society became more able to manipulate the environment and its resources with the aid of its technological prowess, and Western culture began to develop particular attitudes to certain smells, such as body odours and the smells of animals, and olfactory practices, so a gradual differentiation and classification of smells emerges. Certain smells become associated with disease (see Engen 1982), others with foreign places and the exotic, and others with the carnal pleasures and sin. In time, certain smells became associated with the masses and poverty, whilst exotic perfumes and more subtle odours became associated with the rich elite. Eventually, and especially in the twentieth century, more attention was paid to the olfactory as well as visual dimension of products and places. Manufacturers began to add odours to all manner of products, from detergent cleaners to processed foods, because of their established associations and to promote sales of their products. Urban planners also began to take greater note of the olfactory impact of industry, especially of the chemical industry, and steps were taken to encourage siting of such industries down-wind of residential areas. This was, however, part of a wider concern about pollution.

SOUND AND NOISE

There are perhaps three features characteristic of contemporary Western auditory geographies:

1   certain sounds are given *privileged* place over other sounds;
2   an increasing obsession with ever louder sounds, that is sonic *amplification*;
3   interest in recording sounds and *controlling* them.

Our contemporary approach to sound is, it seems, to identify specific sounds or groups of sounds, to copy or synthesise them and/or record and replay them, often amplifying them to make them 'clearer', and above all to control the auditory world and assert power. In Western cultures we seem to dominate space with sound rather than define it by sound (Schafer 1977, 1985). Technologies of recording and amplification play a key role in our manipulation of auditory geographies.

The distinction between unnecessary noise and acceptable, or taken-for-granted loud sounds in the everyday environment is largely determined by convention or political power (see Tempest 1985; Kryter 1985). The sounds of nature seem at times to have been relegated to a background, decorative rather than functional. Inside our buildings and cars, and even public squares and parks, we hide away from natural sounds by playing radios, televisions and compact disc players. In the city, the streets are full of the noise of traffic and the noise of machines dominates the work-place, whether it is on the farm, in the office or in the factory. Human sounds, intentional and unintentional, blanket out much of the auditory world around us. Much of auditory orientation is thus lost and perhaps it is no wonder that the eye gains such dominance over the ear in so many situations of everyday life. The Pelican crossing may bleep, but most people are probably waiting to see the green man flash up.

In the history of Western culture, certain sounds have come to dominate specific soundscapes in particular periods. These human sounds dominate space and directly or indirectly influence the activity of people in those spaces. Schafer (1977, 1985) introduces the theory of 'sacred noise'. He argues that specific sounds are privileged by a society and allowed to dominate its acoustic space. In different historical periods, different sounds have gained this status. In medieval Europe, the church bell had such dominance, demarcating the parish by its reach across the landscape and expressing the power of the church over the people. In the nineteenth century, the steam engine and factory hooter gained dominance, the first symbolising progress and industry, the second giving a new temporal order to people's day-to-day lives. The factory hooter commanded workers to work and signalled the end of the working day. The school bell functioned similarly to punctuate the school day, lesson breaks and playtimes all marked by the bell. Today, the aircraft perhaps acts as a soundmark (see Chapter 6) in our auditory geography and the noise of traffic on our city streets and country roads. Yet, whilst the sounds of aircraft and road traffic are debated as issues of 'noise pollution', the ubiquitous ring of the telephone calls us to attention, demands our immediate action and thus is privileged in our auditory experience. The telephone is a privileged noise, one expressing the power of one individual over another, and is the modern equivalent of the sacred noise of the church bell, now long since drowned in the roar of the traffic.

These privileged sounds are often quite distinctive sounds, loud and somewhat harsh. They call to attention and demand action. They do not form part of a continuous background, but rather are punctuations in experience, structuring space–time patterns of everyday life. Other sounds are dominant in a much less intentional way, accepted often as an inevitable consequence of economic activity and ways of life. Sometimes they are called 'noise pollution', such as traffic noise and aircraft noise, and much research has been done on the side-effects of such auditory geographies on individuals and communities (Berglund *et al.* 1975; Cohen *et al.* 1973, 1981; Korte and Grant 1980; Green 1982). These polluting noises are generally more continuous and monotonous, often loud and associated with machines, and characteristically blanket out other sounds. Such sounds dominate a sound-scape and may also be associated with other effects, such as vibration of buildings and atmospheric pollution (tactile and olfactory effects, respectively) (see Porteous and Mastin 1985).

In addition to sacred noise and noise pollution, one can also identify another type of auditory experience in contemporary Western culture: 'play-back sound'. This is broadcast and recorded sounds – the 24-hour radio stations left on as background in public places, factories and, for much of the day, in homes, and the Walkman worn by so many people today when travelling by plane or train, or when jogging. Play-back sound is chosen for its pleasantness, it provides a context or atmosphere for everyday activities. It is predominantly music and speech, listened to both actively and passively, and played for entertainment and information. Whether radio with news and advertisements, as well as music and chat, or recordings of music on the home hi-fi or Walkman, a key characteristic of these sounds is their electronic recording and reproduction. They are synthesised sounds, equivalent to the photographs and film we substitute for actually being at a place to see it with our own eyes. As a recording, the play-back sound is limited by the technology which is employed, as is graphically illustrated by the differences between 78s, LPs, audiotape and compact discs, and by the common experience of not recognising a familiar voice when first heard on the telephone or a recording. The technological medium transforms the sound and separates it from its immediate context (which may provide other useful non-auditory clues). It reduces it to a certain frequency range and alters the character of the sound. Just as the graphic equaliser can alter the experience of a familiar record, so also we must remember that the digital recording technology is in the end a sampling and enhancing of key features of a sound. A decision has been made which is creative and the play-back sound is a representation, or more accurately a simulation, of the original sound.

Play-back sounds, because they are recorded and can be played again, are often repeatedly heard. Repetition and increased familiarity may give us a sense of security with this auditory geography, but it also disguises how those sounds become important reference points in our auditory experience. Emotions are attached to particular sounds, or tunes, and certain voices attract our attention, and just as with the factory hooter and the church bell, our behaviour in space and time can be affected by the patterns and timings of auditory events on the radio.

Perhaps as important as recording sound, and with a longer history, is amplification of sound. In Western cultures, this has been a dominant dimension of auditory experience. Through amplification, humans can gain access to hidden worlds which the unaided ear cannot reach and also with amplification one can assert a certain power over spaces, as in the open-air rock concert. Amplification once made use of the features of the natural environment and the special qualities of certain sounds, for instance, in the communication between mountain communities by yodelling, or tribal communication in Africa with drums. Various kinds of horn and bell were developed to broadcast messages, such as the church bell, the fog horn and the car horn. And today, electronic amplification is commonly used to assist speakers at conferences, for public address systems in railway stations and airports, and in shops to attract attention to sales and special offers. Amplification aims to spread a specific sound over a greater distance, to make it heard.

The increasing use of amplification may reflect the increasing noise in our environment, but it also reflects an assumption that louder sound is somehow better or more effective. The emphasis is not on listening more carefully, or adapting the type of sound to suit the situation, but simply making it 'bigger', that is, louder. The invention of the hearing-aid also seems to illustrate this definition of sound as loudness. If someone is hard of hearing, we give them a device which amplifies the sound, yet much of hearing impairment is not a loss of absolute hearing but a loss of discernment and differentiation of individual sounds (Mindel and Vernon 1987). Loudness is only one property of sound. The reduction, or abstraction, of auditory experiences to a geography of loudness – as seen in noise pollution and play-back sound especially – is to ignore or underestimate the importance of its other qualities to geographical experience. The timbre of sound, its relationship to the acoustical properties of the environment, its duration and its relationship to other sounds, are all important properties of auditory experience which a loudness geography omits.

Schafer suggests that 'the territorial conquest of space by sound is the expression of visual rather than aural thinking' (1985: 92). The use of amplification and the tendency to see noise as loudness, are symptoms of our concept of auditory experience. In contemporary Western culture we seem to have reassigned sound to a packaged existence, a kind of colouring or decoration to spaces, as in the musak of restaurants and shops and the use of recorded sounds in the home, and elsewhere to use sounds specifically to warn, as in the pelican crossing, supermarket check-out and emergency services. This is an auditory geography which is made up of sounds which are distinctive, often quite loud, and standing out from a general background of less differentiated sounds. We seem less inclined to utilise our ears to listen to the soundscape unfolding about us and to orient ourselves in an auditory world (except if blind perhaps), but instead experience a less immediate auditory geography. This seems to consist of two types of experience: a background of music, speech and noise which often blankets out the fine details of auditory spaces (and natural sounds especially) and sudden, often loud, encounters with warning sounds. This is an auditory geography of hearing rather than listening, or juxtaposition rather than relationship. Much of it is characterised by human generated sounds, many of which are synthesised and repeated incessantly.

9

IMAGES, THEMES AND POSTMODERN GEOGRAPHIES

POSTMODERN GEOGRAPHIES

The term 'postmodern geography' is extremely ambiguous and seems to have many different associations within geography (Dear 1988; Folch-Serra 1989; Harvey 1989; Soja 1989; Curry 1991). It is used here to refer to the experience of contemporary environments and to suggest that a certain style of sensuous experience may be emerging. Visual experience is taken as an example because it seems that the visual has become especially central to the postmodern experience (Baudrillard 1988, Eco 1986). However, this is not meant to exclude the other senses – touch, smell and hearing – nor to dismiss their part in postmodernity. Space does not permit these to be considered in depth here. These other postmodern geographies, and perhaps a more detailed study of the multisensuality of contemporary geographical experience, form an agenda for future research.

In the transformation of sense(s), the visual has been most studied. In contemporary Western culture it is quite evident that the visual is a prominent dimension of day-to-day sensuous experience and visual geographies take many forms (see Chapter 7). Landscape and map, which were explored earlier, each illustrate an apparent abstraction of the visual. This is abstraction in the sense of a reduction of the visual experience to exclusively visual symbols – in the case of landscape, excluding the visual from its nesting within the matrix of other sensuous experiences and establishing a practice of detached contemplation, and the transition, in the case of maps, from picture-like representations to schematic representations, a system of signs – and abstraction in the sense of detachment of visual worlds (or representations) from the actual visible world – the breakage of the sign–reference link, the replacement of representation by resemblance; that is, an order of simulation. 'The effects of technology do not occur at the level of opinions and concepts, but alter sense ratios or patterns of perception steadily and without any resistance' (Baudrillard 1990a: 89).

The transformation of sight and the generation of culturally specific visual geographies in Western culture continually since the Renaissance, might be described as a kind of hyper-realisation of vision. In contemporary society, it is grounded in a culture of consumerism and mass media, and mediated by advanced communication and information technologies. This is a process of transformation in the realm of geographical experience (person–environment encounter, location and spatial organisation, place character) whereby the visual has become effectively hegemonic – a dictatorship of the culturally constituted eye – and technologically mediated 'images' (or 'themes') substitute or replace actual multisensual dialogue with a physical environment. This transformation is most evident in the enclosed and controlled environments of theme-parks and the virtual geographies generated by film and television, but even outside these 'spaces' one can observe a kind of postmodernisation of space, as in the protection of heritage environments such as Albert Dock, Liverpool and The Shambles, York, in which the visual is prominent in the content and coordination of sensuous geographies. Baudrillard's (1983a) orders of simulacra, that is, styles of signification, seem to offer at least a description of the emergence of this contemporary visual geography (see Chapter 10).

This section explores the reconstruction of the visual in the context of film, leisure and retail environments and, briefly, heritage areas. The transformation processes of symbolisation, association, abstraction and reassignment are also evident in the emergence of these contemporary visually dominated geographies, but here a broader set of features which more specifically demarcate the transformation (or hyper-realisation) of vision are identified. These recurring features seem to characterise a hyper-visible world, the paradox of which is increasingly apparent: all is visible yet strangely hidden from view, that is, the hyper-real alienates the individual from complete (and free) participation in a geographical world. (The socio-political critique of postmodernity is not discussed here. What is disguised in postmodernity are important socio-economic hierarchies between producer and consumer.)

LANDSCAPES OF FILM AND TELEVISION

The landscapes of film and television derive from actual environments but when experienced by the viewer become a kind of virtual geography. Like the study of painting or architecture, an analysis of film reveals insights into the way a society at a given point in time constitutes the visual. In any case, for many people television, in particular, forms an important part of the day-to-day routine and the images and ideas presented by television may have much influence on their geographical understanding. In this sense, film is a simulated experience which substitutes for the actual experience. The visual constitution of the film cannot be reduced merely to a technological question since film is dependent on an audience and is produced in a particular social milieu, that is, film is also a cultural product. As with the other arts, film reflects society but also can offer alternative modes of signification which may or may not be adopted by (or influence) society – a complex dialectic. Particularly influential have been the realist British films of the period 1958–63 which related working-class dramas to 'realistic' northern urban-industrial landscapes (e.g. Higson 1984) and contemporary urban and rural based television soaps set in the north of England (e.g. Shields 1991). Despite important differences, both genres illustrate certain visual strategies characteristic of the reconstruction of the visual in contemporary geographical experience.

Geographical images – landscapes constituted by a synthesis of camera shots, of different focus and angle – play a key role in these films. Different actual geographical locations and often also studio scenes, are juxtaposed to create imagined but 'realistic' landscapes. This is as much evident in the realist films (the 'Kitchen Sink' movies), such as *A Taste of Honey, The Loneliness of the Long Distance Runner,* as the later television series, such as *Coronation Street, Last of the Summer Wine, All Creatures Great and Small* (Herriot). The landscape, streets or fields, are not mere background but participate in the drama and add authenticity to it. The landscape becomes a 'sign of reality' (Higson 1984: 9). However, places or landscapes are primarily constituted not as specific and unique but as representative and above all archetypical. Coronation Street is a working-class street, Holmfirth is a Yorkshire village, the 'distant view from the hill of our town' is for Manchester, Sheffield, Huddersfield or some other northern industrial town. These film landscapes are not only synthesised and archetypical, they are also aesthetically cleansed and poetically presented. The film picture bears little relationship to family photographs or local postcards because the audiovisual experience of the professionally crafted film is more 'real' than this: it is presented as art, presented for a spectator.

The spectator's experience of the film landscape is not participatory in the sense of everyday geographical experience, for it is directed by the camera and the intention of the producer. The spectator, or viewer, takes up a relatively passive stance for the images are presented in a continuous series as part of a drama or story. Seeing becomes a consumer activity. In the experience of film landscapes the reciprocity of vision is broken, one can see the characters and places but cannot be seen by them. Film is a peep show and not a dialogue. In actual geographical experience, there is always a possibility of being seen as well as seeing, reciprocity is part of participation. In film, only the film is visible, the spectator is invisible (and the production process behind the film also). The reciprocity is broken. A reciprocity of vision in everyday life means that to see can also open the possibility of being seen, that is visibility. This has been considered in terms of landscape perception (Appleton 1975), the experience of the blind (Hull 1990), and in photography (Berger 1972, 1980) (see Chapter 7).

The landscape images of these films and television series are also iconographic (see Cosgrove and Daniels 1989). In *Coronation Street*, 'the street becomes a symbol and icon of community and of belonging', not just a realist evocation but also 'a prescriptive utopian fantasy' (Shields 1991: 225). Equally, the countryside and village scenes of Yorkshire rural soaps present a series of images symbolic of particular lifestyles and attitudes to social and economic life. These visual images present dramas of a psychological and emotional kind; not social and economic struggles of the political kind. The realist films of the late 1950s and early 1960s equally aestheticised the working-class condition and reinforced stereotypes of working-class folk and the industrial north (Shields 1991; Higson 1984). 'The aesthetic and the psychological block access to the social and historical' (Higson 1984: 11). The drama and the characters are not just situated in a landscape as if set against a background, but the landscape is part of them, and so forms part of their definition and development. The fascination with the visual image becomes a crucial part of the experience of the film. In this sense, the consumption of film landscapes is hedonistic rather than critical, personal rather than social. The reduction to visual images de-politicises and generates a fantasy of 'the real' detached from the everyday experience of the viewers, yet it is one in which they can momentarily 'participate' or watch by switching on (and off) the television or attending the cinema.

The synthesised landscapes of film are iconographic in the sense that they reconstitute actual landscape elements into archetypical images. The reference to those images is within a cultural tradition – in this case, stereotypical images of the north of England – and in the film (or its genre tradition) an agreed code or repertoire of visual images and their interpretation is established, repeated and reinforced. In these realist films and more recent soaps, landscapes, places and geographical features – that is the photographs of them – are not employed as representations but are composed as simulations (in Baudrillard's sense of this term). Furthermore, the iconography is self-referential, that is, more dependent on this cultural tradition than any reference to an original reality, since any number of similar landscapes could be substituted. The realism of northern soaps is tested against other films set in the north – the expectations of viewers – as much as against actual landscapes. Considered alternatively, the film tradition conditions how later producers and consumers (makers and viewers) visually experience actual landscapes and other film presentations (see also Friedberg 1993).

The constitution of the visual is, however, not merely the simple reduction to the iconographic image or an abstraction from everyday reciprocal visual experience. Film also illustrates another distinctive feature of the visual reconstruction of contemporary society, the articulation of a vital tension between the telling of a story (story-line or narrative) and the evocation of place (visual fascination or spectacle) (for a wider discussion see Debord 1983). Higson (1984) explores this narrative/spectacle tension in the 'kitchen-sink' realist films. He also notes related tensions specific to these visual images: between the drabness of the setting and the poetic quality of the landscapes, between documentary realism and romantic atmosphere, and between problem (working-class story) and spectacle (drama, place evocation).

> All novelistic forms have to accommodate both narration and description, both narrative movement to a new and different space and time, and the repetitive description of a single moment in space and time. In cinema the image can both narrate and describe at the same time, but there is still a tension for, although the narrative system struggles to fix the meaning of an image, there is always more than the narrative can hold.
>
> (Higson 1984: 3)

As Heath observes: 'narrative never exhausts the image ... narrative can never contain the whole film which permanently exceeds its own fictions' (1981: 10). In the realist films, the landscapes participate in the story or narrative as 'metaphors for the state of mind of protagonists' (Higson 1984: 3). In the Yorkshire-based television series (*All Creatures Great and Small, Last of the Summer Wine*), the village and countryside scenes, through different seasons and weather conditions, equally – though differently – form part of a story-line. Yet at the same time, the landscapes in film and television are also spectacles as well, evocative of particular types of place (and time) and visually fascinating in themselves. In the realist films, the long-view landscape shots cut across the narrative flow and 'function as spectacle, as a visually pleasurable lure to the spectator's eye' (Higson 1984: 3), and much the same experience of a whole world expressed in a particular style of landscape photography is also present in the spectacle of the Yorkshire Dales in the more recent television series.

This narrative/spectacle distinction and articulation is an important element in these visual geographies and is to be found transposed and equally articulated in the leisure and retail environments of themed landscapes, such as those created by Disney (Bachmann 1990; Wakefield 1990). Higson (1984) characterises the narrative as something lacking, installing a desire to explore, to find out what is missing, to move on to the next scene and the possibility of achieving what is desired. Narrative is perhaps motivated by voyeuristic curiosity. In contrast, the spectator confronted by an image which is so fascinating that it seems complete – where there is no longer any desire to move on, no longer a sense of something lacking – experiences spectacle. Voyeurism is blocked in a moment of fetishism (Higson 1984). Higson emphasises the contradiction of these elements in the visual construction of the realist films, yet closer analysis suggests that whilst still pulling in opposing directions, narrative and spectacle are each dependent on the other for their vitality and continuity. The narrative needs the visual pleasure of the spectacle and its rich evocation of place, yet if the spectacle is the only focus it soon loses its rationale without the narrative and the fascination fades. The articulation between the two elements enhances them both and maintains the 'realism' of the visual surface, the landscapes of the film. Perhaps, even, the success of the hyper-reality of the film landscapes – more real than real, rooted in culturally defined archetypes – lies in the necessary tension of this partnership. Wollen (1980: 25) expresses this slightly differently:

> place implies memory, reverie ... the imaginary ... Place also implies displacement, being elsewhere, being a stranger. Films are like imaginary journeys; the cinema is a magic means of transport to distant places. Places are functions of narrative (actions must take place somewhere) yet the fascination of film is often in the places themselves.

The sense of being elsewhere is also recognised by Hopkins (1990) in his study of West Edmonton Mall. Here he identifies a property of 'elsewhereness' (see next section). The narrative/spectacle articulation is therefore evident not only in films but also the controlled environments of theme-parks, shopping malls and even heritage areas which each have a kind of 'story' associated with them – cartoon adventures at Disney theme-parks, lifestyle images in malls, and reconstructed pasts in heritage areas. Yet these landscapes, or places, are aesthetic experiences in themselves at the same time, and necessarily so.

THE SENSUOUS EXPERIENCE OF THEMESCAPES

Shopping malls, leisure parks (sometimes called theme-parks) and heritage areas might be described as 'themescapes' or themed environments. The ways in which visual experience is constituted are particularly important in these built environments. Each, to varying degrees or in different manifestations, parallels the experience of film. The degree to which they share the visual experience of the film lies in part on the degree of enclosedness of the environment. The totally enclosed and controlled environments of the Disney theme-park (Eco 1986; Wakefield 1990) or the themed shopping mall (Hopkins 1990; Chaney 1990) especially repeat the same characteristics. It also reflects the relationship between the design of these environments and traditions of film and television. The history and design of shopping malls provides important insights into the evolution of shopping practices and the emergence of hyper-real geographies of contemporary retail environments (see Northern and Haskall 1977; Maitland 1985).

The more open and less comprehensively managed environments of older cities, where old buildings and other features of past urbanism are protected and maintained (heritage areas) by planning controls and financial support, and the introduction of design features to such areas and other parts of the urban fabric to enhance a revival of historical urban style (e.g. Albert Dock, Liverpool; Little London, Bradford; The Shambles, York) emphasise the visual appearance of that heritage in a sanitised and renovated form, and repeat similar characteristics but with inevitably more intrusion from alternative experiences (Hewison 1989). Nevertheless, this visualism in heritage areas is not just architectural, it can also include the encouragement of traditional types of shop and the revival of street music (as in the past), and even the 'staging' regularly of period dress by shopkeepers keen to participate in the fantasy of a past (e.g. Halifax, West Yorkshire). The Shambles, York, may take on the heritage theme but it is not so complete as the controlled environments of the re-creation of Viking York in the Jorvik Heritage Experience. The latter is one of a number of time-travel museum experiences which have developed throughout Britain and America. Like the traditional indoor museum, this is an enclosed environment, but it is not just a visual display. The Jorvik Viking experience seeks to excite a fuller sensuous encounter with the past through the use of tactile, olfactory and auditory dimensions to the experience as well as the visual re-creation. Such experiences go beyond the film spectacle towards the virtual reality of a simulator.

The 'actual' environments of leisure and retailing are defined by themes. These themes are predominantly visual in presentation and articulation. They are surface textures; a kind of packaging. The themes are manipulated in ways similar to the landscape images in films. Where the camera samples actual landscapes and recomposes them into archetypical landscapes, the theme designer (with the exception of heritage areas perhaps) constructs a resemblance, or simulation, of a particular place or, more accurately described, style of place. The choice of themes and presentation of them is grounded in a cultural tradition and much reference is made to television and cinema images and stereotypes. This cross-referencing between cultural media is especially evident in leisure parks and shopping malls (Wakefield 1990).

The 'theme' is not dissimilar to the 'image' as constituted in contemporary film. The themed environment is primarily constituted visually, being first a surface and architecturally associated with facadism and stylistic gestures. It is perhaps useful to distinguish between:

1   'Themescape' – a themed environment, and specifically a space or place which is identified by a single coherent theme or idea. Frontierland, in Disneyworld, would be an example of such a themed environment. Such themescapes can be quite varied but ultimately hold to an overarching theme.
2   'Theme environments' – a collection of themescapes, such as Disneyworld's central area of Mainstreet, Frontierland, Fantasyland, Tommorrowland, Adventureland. The theme environment is an aggregate surface

165

made up of distinct areas juxtaposed but not dependent on any specific relationship to one another – a collage landscape.

The term 'themed environment' specifically refers to one geographical space and is therefore usefully replaced by the term themescape. The '-scape' emphasises a geographical space defined in terms of its visual composition and coherent wholeness. Technical definitions for a 'themescape' are as yet still quite illusive and when a coherent (stylistically) space becomes a themed one is far from definitive. However, themescapes can be recognised by their vivid visual appeal, with strong and coherent references to particular places and periods elsewhere (Hopkins 1990). The theme is not a replica but rather a resemblance, equivalence rather than copy. It is rooted in a stereotypical image and adapted to the practical demands of the current environment to be themed. The street in Disneyland themed as a nineteenth-century European Boulevard is archetypical, not specific, and proliferates with references to stereotypical images from various European cities. Further, the shopping mall whilst clothed in a theme is nevertheless occupied with current retail outlets – the same old stores of modern USA (or Britain, etc.), and perhaps discreetly, the ubiquitous credit card signs. The Hawaiian Beach complex has palm trees and warm waves but with its chutes and diving boards it is still a leisure pool.

The themes are surprisingly limited in range and many of the 'old favourites' recur at shopping mall after shopping mall, such as references to images of the past (the Wild West, the European Street), exotic locations (famous tourist spots), childhood fantasies and futuristic images, and also images from film and television. The success of a theme is grounded in reinforcing widely shared place stereotypes and dreams. The economic effectiveness of themescapes is ultimately a question of effective market research and the translation of this into appropriate theme designs. Without mass travel and the mass media, especially television, much of the themescape culture would perhaps be lacking. As with the film image, discussed earlier, the theme environment is not just a spectacle but is also a narrative. The theme is an evocation of a place style and a fascination for the eye. The theme is also a story-line for a particular shopping and/or leisure experience. In the heritage area, identifying a particular theme from the past both enhances an aesthetic experience and narrates a story of the reconstituted past (itself conveniently inclusive of a shopping/tourist experience) which is attractive to visitors and investors alike. The theme is therefore both spectacle and narrative, art and commodity. As Hopkins observes for the mega-mall, 'context itself ... [is] a consumable product' (1990: 11). Jameson expressed a similar conclusion when he argued that the image is 'the final form of commodity reification' (1984: 66).

The creation of themescapes and theme environments is often described as Disneyfication and refers most generally to the spectacular leisure/retail self-contained environments proliferating in North America and increasingly in Europe and elsewhere in the world (Relph 1981; Eco 1986). Theme environments often juxtapose radically different themescapes – that is spaces of a coherent theme – across a given landscape, offering a kind of place collage rather than smooth variations in place references. This characteristic perhaps echoes the television experience more than cinema and reinforces the sense of fantasy. The themescape does not pretend to be anything other than escapism. It is a celebratory fake and deals in resemblances only. This strategy is rooted in the original Disney conception of theme parks, totally enclosed environments of juxtaposed themescapes. Further, the theme is inclusive not only of building surfaces but also of people (actors, shop assistants) in them and even in some cases the products and services sold. The Disney retailing outlet typifies this extension of the visual image of themed environments into purchases.

Therefore, theme environments seem to be rooted in two important characteristics of postmodern culture, both centred on the reconstruction of the visual. First, the origins of theme techniques seem to lie in film-making and this is illustrated in the path-breaking innovations of the Disney theme-parks (Wakefield 1990; Bachmann 1990; Rojek 1993). As already discussed this film heritage introduces the articulation of visual narrative and visual spectacle in geographical space.

Second, the theme environment is rooted in visual abstraction and the wider Western experience of aesthetic abstraction. The contemporary experience separates the act of creation (art) from the act of appreciation (aesthetics), valuing the product over the process. Art is reduced to technique and craftsmanship to manufacture, whilst the art work or object becomes spectacle and enters the economy of the commodity (see Rojek 1993). In contrast, in aboriginal cultures, such as the Inuit, the process of artistic creation was valued far more than the product and, historically, once completed was discarded (Carpenter 1973). Art for the aboriginal was a relationship with the environment, a communication with living things and forces and forms of nature, and establishing an understanding of its intricate features. In theme environments, the image or theme is a product to be consumed, like the wares in the shops, and landscape experience is a consumer activity. Therefore, the producer and consumer, the active and the passive, are separated and hierarchically related, as much in the film and the themescape as in other commodities.

Together, these two ideas: narration/spectacle articulation and commodification/consumerism suggest an important socio-economic rationale for the visual reconstruction of geographical experience: the reinforcing of capitalist social relations. Theme environments are profit-generating machines, using hypervisibility to make invisible (or less visible) traditional practical relationships (see Hopkins 1990).

As with film, the themescapes are iconographic. Hopkins (1990) recognises both metonymical icons and placial (place-specific) icons in the

167

shopping mall environment. An icon is a type of sign rooted in social convention. The theme environment thus becomes a system of signs. As Berger notes: 'when the relationship or "syntax" between a signifier and a signified is one of resemblance as in the case of the photograph [and the themescape] the sign is "iconic"' (Berger 1984: 12). Hopkins argues that placial icons 'attempt to simulate the characteristics and uses of other places; they are "spatial metaphors" in that they act as substitutes for the original referent or place' (1990: 4). Metonymical icons do not act as placial substitutes but refer to a larger whole through their associate characteristics. Placial icons supersede metonymical ones. The visual resemblances are presented not literally but 'aesthetically'. For instance, scale may be reduced or expanded for particular effect, colours adjusted to enhance the spectacle or incorporate a corporate style. Further, various optical illusions are frequently employed. For instance, the upper floors of the Europa Boulevard in West Edmonton Mall are progressively reduced in size to give an illusion of height, a technique reminiscent of Disney's Mainstreet USA. The glass roof and the introduction of mature-looking small trees can all add to the sense of space and even outsideness. However, 'a shopping centre is an enclosed space so that customers, once they have entered, have no experience of the physical environment that is other than completely culturally controlled' (Chaney 1990: 49). The complete theme environment is a visual fantasy akin to film except that one can 'get inside it', experience the spectacle as all surrounding encounter.

Hopkins (1990) identifies a number of myths which are reinforced by the visual spectacle of the mall: egalitarianism through consumption, personal transformation or improvement through consumption, millionaire status, the mall as a community, that is as a public rather than private space, and finally as liberating and escapist. Of course, the theme environment does not transcend socio-economic struggles – shoppers still get into credit card debt, for instance – but hides them. 'Spectacle is not just a collection of images [however], but a social relation among people mediated by images' (Debord 1983: Section 4).

Heritage experiences, such as Jorvik at York, apply the theme-park concept to the museum and achieve simulations comparable to other theme environments. Though ostensibly grounded in a 'real' past, much imagination is used in these re-creations. It is heritage rather than history (O'Rourke 1990; Hewison 1989). O'Rourke observes how the excitement over the new heritage experiences parallels that over the early film industry – a telling observation. The Beamish Open Air museum in the North East of England and the Weald and Downland Museum in Sussex, each try to recreate a 'living experience' of the past by using buildings and artefacts from the past laid out in a parkland area. These open-air museums cannot mask out the passing of a plane overhead or the noise of traffic on a road beyond the site, but at Beamish the effective use of actor/curators to demonstrate activities in the past enhances the realism of the experience. However, such museums tend to juxtapose different elements from the past and different periods in close physical proximity, making a collage of the past not dissimilar to the structure of theme-parks and television.

The heritage or conservation area established in many old cities, and some rural villages, is a more problematic and interesting example. Here, a similar process of theming takes place not through manufacture but through renovation and conservation. The distinction between totally constructed theme environments and a renovated heritage space is increasingly blurred, especially in housing projects involving the conversion of old mills into flats and the addition of similar new structures (to blend in), as in Skipton, North Yorkshire, and Albert Dock, Liverpool. This blurring is increased by the fact that heritage becomes a key theme and new structures are constructed to harmonise with this visual form. Wakefield describes this as 'historical simulation' and goes on to suggest, 'theme parks, far from being the substitute for American false/non history, may well transpire to be the epitaph of European real/actual history' (1990: 113).

THE RECONSTRUCTION OF THE VISUAL

The film, theme-park, shopping mall and heritage area, each offer examples of visual geographies and contexts where the visual is increasingly hege-monic. In other words, the visual and, importantly, a specific style of visualism, is increasingly seen to dominate the geographical experience and mediate, or condition, the rest of our sensuous, emotional and practical (economic) experience in that context or place. This reconstruction of the visual seems to constitute the immediate impact of postmodernism on geographical experience. However, it is more than just a reduction to the visual, it is an obscuring by the visual – hyper-visibility – of significant economic and social relationships (conflicts, cooperation) with a kind of utopian image, a fantasy experience, and increasingly the redefinition of much more than certain enclosed environments but whole landscapes of everyday life or the way we come to perceive the environments beyond the cinema and television, theme-park and mall. The reconstruction of the visual also leads to an associated reconstruction of non-visual experience and therefore of haptic, olfactory and auditory experience and the geographies they constitute. This wider reconstruction of sense and reality constitutes an agenda for future research.

In summary, in this chapter we have identified nine characteristics of the reconstruction of the visual.

1   *Visual enclosure or reductionism* – the enclosure of space in a realm dominated by visual images, facades or surfaces, and articulated by a logic of visual technology.

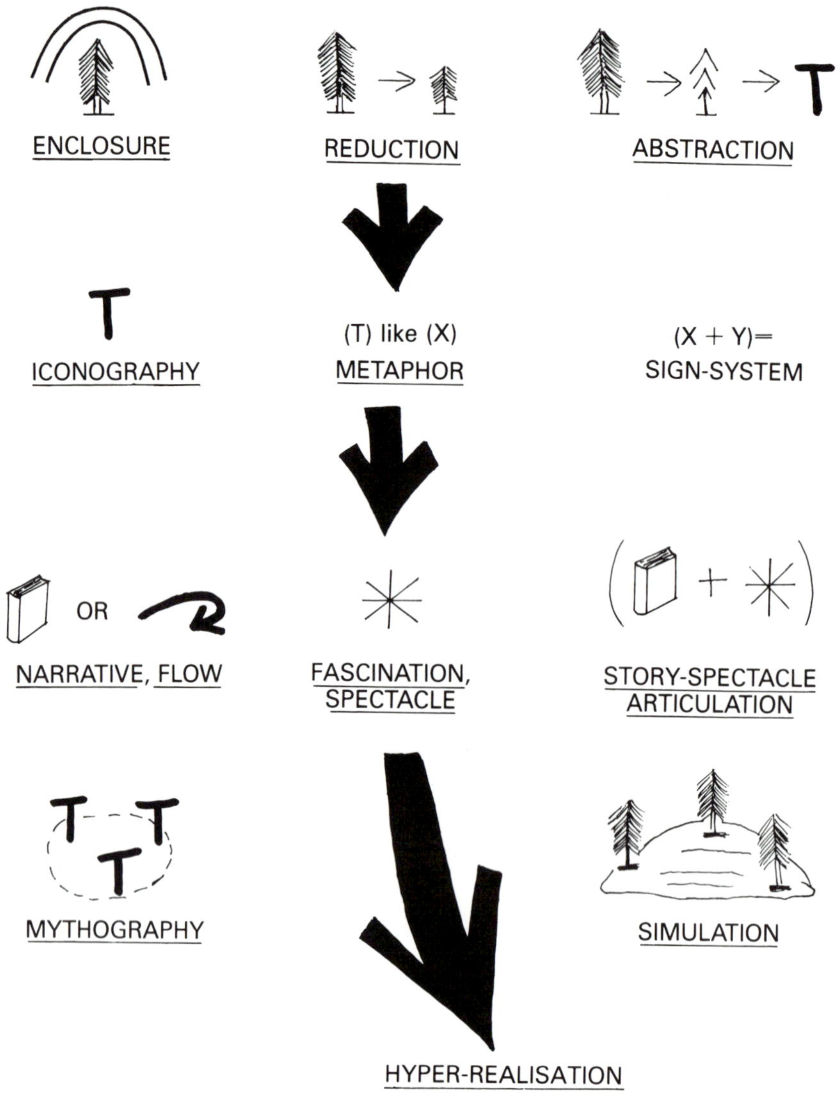

ENCLOSURE

REDUCTION

ABSTRACTION

T

ICONOGRAPHY

(T) like (X)
METAPHOR

(X + Y)=
SIGN-SYSTEM

NARRATIVE, FLOW

FASCINATION,
SPECTACLE

STORY-SPECTACLE
ARTICULATION

MYTHOGRAPHY

SIMULATION

HYPER-REALISATION

Figure 9.1 Reconstruction of sensuous geographies: a hypothesis

2   *Abstraction of the visual* – the breakage of the reciprocity of the visual and the maintenance of an aesthetic abstraction.

3   *Visual iconography* – the image or theme expressed as a system of signs, rooted in a cultural tradition dominated by the images of the mass visual media (notably television).

4   *The visual as simulation* – the visual as resemblance rather than representation or copy, and essentially of the order of simulation.

5   *Visual metaphor* – the reference to other places and times through the visual presentations.

6   *Visual narrative* – space having a visually articulated story-line akin to film narrative.

7   *Visual fascination* – place evocations that fascinate the eye, a spectacle in itself.

8   *Story/Spectacle articulation* – the visual narrative and spectacle both contradict and reinforce one another in the illusion of the 'reality' which constitutes a hyper-real environment and conspire to hide the actual socio-economic relations of producer and consumer.

9   *Visual mythography* – the consumer/spectator utopia which the reconstruction of the visual – in total – generates and reinforces.

The reconstruction of the visual, whilst tending to be almost hegemonic, is also evident increasingly in the other sense modes and perhaps suggests a hyper-realisation of geographical experience in general. To suggest the hegemony of a specific style of visual experience is not to exclude the other senses, but to perhaps suggest that they are increasingly defined in relationship to the visual and its associate technologies in the contexts discussed here – film image and themescape. The transfer of a visual logic onto the non-visual reduces other sensuous geographies to visual-like ones and so hides their distinctive contribution to the multisensual nature of geographical experience and understanding.

As a preliminary to formulating future research agendas, each of these elements of visual reconstruction can, perhaps, be identified to a certain degree, or in similar processes, in the other sense dimensions – haptic, olfactory and auditory (Figure 9.1).

171

10

SENSE AND HYPER-REALITY

TOWARDS A THEORY OF SENSUOUS GEOGRAPHY

A theory of sensuous geography could seek to explain the relationship between sense and reality, and specifically account for the transformation of the senses and sensuous geographies in Western culture. Such a theory cannot be a positive or absolute theory since the transformation of the senses is both complex and on-going. What such a theory of sensuous geography can offer is a normative description or framework of possible pathways of transformation and potential changes in the matrix of sense and reality. This normative theory provides a kind of lattice, constituted of socio-historical transformations, grounded in the physical limits of human-ness and the environment, and defined and re-defined by the ingenuity of human invention (technology) and changing cultural practices and values, all operating over time, in complex networks of contingent relationships and situated in specific places.

This theoretical strategy has roots in Foucault's dispositif (summarised in Shields 1991: 43–45) and Bakhtin's chronotopical analysis (summarised by Folch-Serra 1990), and the ideas of Baudrillard (e.g. 1983a). Here, a preliminary sketch of this theory is made, concentrating on an application of Baudrillard's ideas about the emergence of hyper-reality. This contemporary experience – the postmodern space or hyper-reality – is illustrated most immediately by the sensuous geographies of cinema and television and the themescapes of leisure parks (such as those of Disney) and shopping malls (see Chapter 9) but this hyper-real experience extends far beyond these spaces to include a re-definition of much else of our geographical experience (Eco 1986). In the transformations of haptic, olfactory and auditory experience in contemporary society, socially and technologically, individually and in relationship to one another and to the visual, one can also identify similar processes of hyper-realisation (see Chapter 8).

The 'hyper-realisation' of sense and the senses might be defined as the separation out of individual sense modes, the reduction or simplification of these sense modes to specific features or roles, the organisation of the senses hierarchically, and the assignment of the lesser senses, thus defined, under the hegemony of one re-defined sense. This transformation of the senses involves the processes of symbolisation, association, abstraction and reassignment. It is a transition from multisensual and complex 'natural' or direct sensuous experiences to single sense and simplified 'synthetic' and simulated experiences. The 'reality' which the senses give access to is also transformed by the same cultural and technological changes which have stimulated, perhaps even necessitated, the hyper-realisation of the senses. In other words, this transformation affects sense both as sensation and meaning. It is a reduction and simplification of reality akin to a kind of domestication of that reality, in which human control over the physical, mental, social and cultural world becomes ever more complete and human experience – including geographical experience – is increasingly mediated by technologies and cultural practices, and 'reality' becomes effectively that which is synthesised and manufactured by human design.

This is a far cry from the world of the creator God of religion, or the world of objective facts obtained by observation with the senses and analysed by science. This is a new order of sense and reality, a reality that is designed, marketed and consumed perpetually. It is a reality without original reference, it is self-referential. This is the hyper-reality of Baudrillard, a reality more real than real, where the senses are subservient to the dictates of desire and a continuous unfulfilled hedonism. It is not a reality like the existential world of existence before essence or of the individual creative self (Samuels 1978; Kaufman 1975). It is a reality of hidden power relations, a realm of hierarchical dichotomies between the designers and manufacturers of simulations and the consumers or spectators of them (Wakefield 1990; Hopkins 1990; Rojek 1993). This postmodern reality – defined here in the abstract extreme – reduces the senses to limited salient characteristics and submits them to the hyper-stimulation of synthesised sensuous encounters.

This chapter is deliberately speculative. It draws upon the writings of Jean Baudrillard and specifically his concepts of hyper-reality and the order of simulation. Whilst his writings are not directly geographical, many of his observations draw upon his attempt to explain contemporary social and environmental experiences (e.g. Baudrillard 1988).

ORDERS OF SIMULACRA

Baudrillard is a fascinating writer and especially relevant for social and cultural geography (Baudrillard 1983a, 1983b, 1988, 1990a, 1990b, 1992). His key works discuss the experience of contemporary society and environment, the so-called *postmodern experience*, and have been much discussed in the social sciences, including geography (e.g. Wakefield 1990; Poster 1990; Clarke 1992). His ideas about the nature of reality, including the concept of 'hyper-reality', and styles of representation are particularly relevant to an analysis of contemporary sensuous geographies (such as the 'themescapes') and in explicating the changing cultural definition of the senses and the relationship of sense and reality.

The 'orders of simulacra' represent, in effect, Baudrillard's theory of signification, that is, the manifestations of different styles of representation and ultimately of sense and reality shared by a community (Baudrillard 1983a). Each of these orders seems to correlate with changes in sensuous geographies, especially those experienced in the emergence of contemporary, high-technology, mass media, consumer cultures of the West. Here, Baudrillard's orders of simulacra are only briefly sketched and offered as part of a tentative theory of sensuous geographies, that is, an explanation of the changing nature of sensuous experience and the 'reality' which is constituted by the senses.

Baudrillard, in typically bold style, argues that

> abstraction today is no longer that of the map, the double, the mirror or the concept. Simulation is no longer that of territory, a referential being or a substance. It is the generation of models of a real without origin or reality: a hyper-real. The territory no longer precedes the map, nor survives it. Henceforth, it is the map that precedes the territory – PRECESSION OF SIMULACRA – it is the map that engenders the territory ...
>
> (Baudrillard 1983a: 2)

Like so many other writers, he tends to take visual examples of representation, words such as 'image' and 'map' constitute a visual geography of his argument, but his ideas have much wider application to haptic, olfactory and auditory geographies as well. Interestingly, in discussing the mummy of Rameses II, Baudrillard makes the telling observation about Western culture: 'we need a visible past, a visible continuum, a visible myth of origin to reassure us of our ends ...' (Baudrillard 1983a: 18–19). In exploring the evolution of the image, Baudrillard observes a

> succession of phases of the image: it is the reflection of a basic reality, it masks and perverts a basic reality, it masks the *absence* of a reality, it bears no relation to any reality whatever: it is its own pure simulacrum.
>
> (1983a: 11)

Baudrillard suggests three orders of 'appearance' or signification (as regards sense and reality) which have followed one another since the Renaissance (1983a: 83):

1   *counterfeit* – is the dominant scheme of the classical period, from the Renaissance to the industrial revolution;
2   *production* – is the dominant scheme of the industrial era;
3   *simulation* – is the dominant scheme of the current information technology age.

Baudrillard presents his orders of simulacra in various guises, each illustrating manifestations of these orders and their inter-relationship. For instance, these orders parallel mutations in the law of value, that is, socio-economic changes in Western society since the Renaissance: the first order is based on the natural law of value, the second on the commerical law of value and the third on a structural law of value. Second, reflecting on science he identifies his three orders as correlate with: a metaphysics of being and appearance, of energy and determination, of indeterminacy and code (1983a: 103). The latter is the world of information technology and digital codes. He argues that: 'the entire system of communication has passed from that of a syntactically complex language structure to a binary sign system of question/answer ...' (1983a: 116–117).

Counterfeit, production and simulation each describe the nature of relationships between a socio-economic system and the world in which it is situated or, humanistically, they describe styles of relationship between person and world, or people and place. Counterfeit arises out of an age – at the start of the Renaissance – when people began to question accepted dogmas of their time and, in particular, the 'truths' enshrined in religion and mythical belief. With the beginnings of a more modern science, the direct data of the senses gained increasing importance. Here the senses offer naive data, basic sensations of the fundamental order of things. That order is an imperfect order, as our senses are not able to discern more than a partial reality and have access only to the copy which is the mortal earth. The perfect original remains in heaven and not directly accessible to the senses. Any representation is therefore a counterfeit, a copy and an imperfect likeness of the original creation. The counterfeit is not due to any corruption of the original, but to our own sensuous limitations.

With the development of modern science and industry, especially from the seventeenth to nineteenth centuries, the role of the senses was elevated. Religion retreated with the onslaught of an objective science grounded in the belief that the senses did give direct access to the world, to a true knowledge of the order of things. In the order of production, therefore, the senses are sources of accurate data about the nature of reality and can be assisted by various inventions, such as telescopes, microscopes, amplifiers and instruments to record sensuous experiences often beyond the immediate reach of our bodily senses. In the order of production, sensuous geography is equivalent to an actual geography, without any reference to a more perfect form, and the senses are extended by technological devices. This is also the era when human societies begin to gain increasing control and confidence in manipulating the physical environment and ecological systems. The first attempts to capture, record and even to synthesise sensuous experience occurs. Also, towards the end of the period, important progress is made in translating (encoding) and interpreting (decoding) sensuous encounters through other media and forms, such as electricity and the now ubiquitous telephone. This early synthetic production is indicative of the next phase.

The current phase, which seems to emerge with the twentieth century, is characterised by the order of simulation. Here, everyday sensuous experience is increasingly mediated by synthetic sensuous media, such as telephones, televisions and computers. So widespread have such mediated experiences become that we have forgotten how and to what extent they mediate, that is, translate and transform our sensuous experience of the world. Whilst we still do experience direct sensuous geographies, increasingly these have been pre-conditioned by our experience of technologically mediated sensuous geographies. In this order of simulation, the reference to an original becomes increasingly irrelevant as the mediated forms refer to themselves and science seeks to make ever more realistic simulations of 'reality'. These simulations are not mere representations, that is, imperfect copies of an original, but rather they are better than any original, clearer and fuller, and so forgetful of originals and sustained by reference to other simulations. This is especially evident in the visual geographies of the media (television and cinema) and the auditory geographies of recording technology (the progression: records, tapes, compact discs, digital audio tape, minidiscs ...). In the order of simulation, the attempt is not to access the original reality more clearly – in this sense the technologies of extension noted earlier are redundant – rather it is to generate or synthesise a reality more real and more perfect, one which fulfils our desires. This is the hyper-real, more real than real (see next section). This postmodern or hyper-real geography is 'a digital space, a magnetic field for the code [of question/ answer cells]' (Baudrillard 1983a: 138). Baudrillard (1983a: 143–144) identifies several modalities of this order of simulation: deconstruction of the real into details flattening linearity and seriality of partial objects; the endless reflected vision, all the games of duplication and reduplication of the object in detail; the property of serial form.

Whilst it is tempting to interpret the orders of simulacra as a kind of chronology – for which there is some degree of evidence in the development of Western culture and science – it is more appropriate to conceive of them as 'styles' which subsist at certain moments in time and place, manifesting themselves in many different ways, usually with more than one form co-existing at any socio-historic moment, but generally with one order having dominance (except perhaps in periods of transition). In this sense, the orders of simulacra can be interpreted as a kind of explanatory lattice, a framework to guide a normative theory of sensuous geography.

HYPER-REALITY AND VIRTUAL GEOGRAPHIES

The hyper-real and virtual geographies are features of the order of simulation; what is sometimes described as the postmodern condition. As we have already demonstrated, the hyper-real is the concept of a reality more real than real, a simulated sensuous geography. Virtual geographies, in this context, refers to the geographical experiences which subsist within, or we experience (or consume) when, watching films and television or on visits to themescapes such as Disneyworld and the Jorvik Museum. These are not encounters with 'reality' in the direct and original meaning of encounter (Seamon 1979) but a technologically and culturally mediated experience, one where we have almost forgotten that this is the case – a simulated sensuous geography. Such virtual geographies are 'real' in so far as they are hyper-real. The study of such geographical experiences is an important direction for future research in sensuous geographies.

In the third order, that of simulation, 'the very definition of the real becomes: *that of which it is possible to give an equivalent reproduction*' (Baudrillard 1983a: 146). The real is also not just that which can be reproduced but that which is always already reproduced. This is hyper-reality, not the end of the real, but, according to Baudrillard, it is the limit of the real. Here, sensuous geography becomes a kind of consumer activity, a consumption of reproducible experiences. Themescapes such as Disney-world become the perfect model of the order of simulation (Baudrillard 1983a: 23–26; Wakefield 1990; Rojek 1993). Disneyland (and Disneyworld) present an imaginary order to make us believe the rest (the 'real' America) is real, when in fact all of Los Angeles and the America that surrounds it participate in the order of the hyper-real and simulation. 'It is no longer a question of a false representation of reality, but of concealing the fact that the real is no longer real ...' (Baudrillard 1983a: 25). The sensuous geographies of themescapes are so hyper-real – more real than real – that they become hegemonic, mediating the experience of environments beyond themselves, transforming the wider world into something less real (see Eco 1986).

Therefore, a number of key characteristics of hyper-real geographies can be identified.

1 *Hyper-sensuous* – in hyper-real geographies, the senses are experienced in the transformed state of hyper-senses. Each sense is reduced to a limited array of features and these are heightened, or exploited, in the generation of an experience more realistic than ever an original experi-ence could be – 'confused' by its range of features, evident and implicative, and the multisensuality of experience. Often, the sensuous experience seems to be dominantly 'orchestrated' by one sense, such as vision, and the experience has great immediate clarity or vividness. This sense of the more real than real is a kind of excess of reality.

2 *Hegemonic* – the hyper-real geography not only is experienced in a specific place (such as a themescape) and at a certain time, but it is so fascinating or captivating in the first encounter and the repeated encounters that it comes to influence, and even dominate, the perception of all other geographical experience in other spaces. In this sense, the

hyper-real experience replaces the possibility of a test against an original by offering itself as the standard, reinforcing its self-referentiality which dismisses the need for any reference to an original (other than itself). Hyperreality spreads beyond the designed or built environments of theme-parks and shopping malls, to encompass attitudes to and experiences of other built environments, including historical places (heritage areas) and even the countryside, and so reconstitutes them as open to hyper-real interpretation.

3   *Consumerist* – geographical experience was once adequately described by words such as 'encounter', 'interaction', 'interdependence' and 'participation'. Each of these identifies a structural relationship involving the acquisition of geographical knowledge and the development of appropriate behaviour with respect to the environment. These relationships are physical, biological, ecological and functional. The hyper-real experience is, however, an economic and cultural relationship. It is an experience constituted of established images and themes, grounded in an accepted cultural tradition, and located within an economy of buying and selling. The hyper-real experience is presented to the individual as a product, a commodity which can be bought and consumed. The hyper-sensuous nature of the experience and its replicability (the reproduction of images and themes) make product definition both possible and necessary. The individual, or consumer, learns to desire particular products (assisted by advertising) and providers of hyper-real geographies seek to satisfy and sustain these desires. Sensuous experience in the hyper-real context is not aesthetic, nor merely functional, but rather it is hedonistic and involves – to a large degree – consumption for its own sake, that is, to maintain or re-experience the ultimate satisfaction of hyper-sensuous experience.

Hyper-real geographies tend to be characterised by surface, a 'packaging' which hides the workings of the technology and economy which enables these experiences to be available. In this sense, hyper-real geographies lack depth – for the consumer – and are virtual in subsisting at the level of an image, theme or gesture. Whilst the visual is most readily identified as an important surface geography in postmodernity, it is possible to identify similar hyper-real sensuous geographies – haptic, olfactory, auditory – and a kind of multisensual hyper-real geography. The virtual reality technologies which seek to simulate sensuous experience of space and encounter with objects in space, such as the flight simulator, can generate complex visual and auditory worlds – and, increasingly, with various instrument and 'data-glove' extensions, tactile experiences. These virtual geographies may become increasingly important in day-to-day experience of the environment, both in the context of leisure, education and training, and in assisting other productive activities. It would therefore seem an important topic for future research in sensuous geographies – how virtual geographies are generated and the way in which the senses are manipulated to generate realistic geographical experiences.

Sensuous geography is, therefore, not just a study of direct encounter with the environment and the role of the senses in orientation, spatial relationships and organisation, and place characterisation. It is also the study of indirect experiences of the environment, both the virtual geographies mediated by technology and the metaphorical (or mental) geographies of literature and art. Further, remembering that sense is both sensation and meaning, and identifying a linkage between sense and reality, suggests the possibility of normative theory of sensuous geography. This theory has been sketched but future research is needed to develop and evaluate it more rigorously.

BIBLIOGRAPHY

Aalders, J. (1980) 'Computer mapping: I want to start ... but how', *The Cartographic Journal* 17 (1): 21–25.

Ackerman, D. (1990) *A Natural History of the Senses*, London: Chapman Publishers.

Aiello, J.R., Baum, A. and Gormley, F.P. (1981) 'Social determinants of residential crowding stress', *Personality* and *Social Psychology Bulletin* 7: 643–649.

Aitken, S. (1991) 'Person–environment theories in contemporary perceptual and behavioural geography: Part 1', *Progress in Human Geography* 15 (2): 79–193.

Allport, F. (1955) *Theories of Perception and the Concept of Structure*, New York: John Wiley.

Amoore, J.E. (1970) *Molecular Basis of Odor*, Springfield, IL: C.C. Thomas.

Appleton, J. (1975) *The Experience of Landscape*, London: John Wiley.

Arakawa, C. (1988) 'Living conditions in public places inside detached houses built by the housing and urban development corporation – on living rooms and space to receive guests', *Journal of Home Economics of Japan*, 39: 725–730 (with English abstract).

Arnot, C. (1991) 'Relax with the soothing essence of mum', *The Independent* 23 July: 13.

Bachmann, C. (1990) 'Le Fantasme urbain des Mickeys', *Urbanisme et Architecture* 234: 66.

Bagrow, L. (1985) *The History of Cartography*, Chicago: Precedent Publishing.

Bakhtin, M. (1986a) *The Dialogical Imagination* (ed. M. Holqvist), Austin, Texas: University of Texas Press.

Bakhtin, M. (1986b) *Speech Genres and Other Late Essays*, Austin, Texas: University of Texas Press.

Baldcock, E. (1966) 'Milestones of mapping', *The Cartographer* 3: 89–102.

Bartley, H. (1972) *Perception in Everyday Life*, New York: Harper & Row.

Bates, H.E. (1969) *The Vanished World*, London: Michael Joseph.

Baudrillard, J. (1983a) *Simulations* (trans. P. Foss, P. Patton, P. Beitchman), New York: Semiotext(e).

Baudrillard, J. (1983b) *In the Shadow of Silent Majorities and Other Essays* (trans. P. Foss, P. Patton, J. Johnson), New York: Semiotext(e).

Baudrillard, J. (1988) *America* (trans. C. Turner), London: Verso.

Baudrillard, J. (1990a) *Revenge of the Crystal* (ed. & trans. P. Foss and J. Pefanis), London: Pluto Press.

Baudrillard, J. (1990b) *Fatal Strategies* (trans. P. Beitchman and W. Niesluchowski), New York: Semiotext(e)/Pluto.

Baudrillard, J. (1992) *Jean Braudrillard: Selected Writings* (ed. M. Poster), Oxford: Polity Press/Blackwell.

Baum, A. and Koman, S. (1976) 'Differential response to anticipated crowding: psychological effects of social and spatial density', *Journal of Personality and Social Psychology* 34: 526–536.

Beach, H. (1988) *The Saami of Lapland*, Minority Rights Group Report 55.

Bekesy, G. von (1964) *Experiments in Hearing* (trans. E. Weaver), New York: McGraw-Hill.

Berger, A. (1984) *Signs in Contemporary Society*, New York: Longman.

Berger, J. (1972) *Ways of Seeing*, London: Pelican Books/Penguin/BBC.

Berger, J. (1980) *About Looking*, London: Writers and Readers Publishing Cooperative.

Berglund, B., Berglund, U. and Lindvall, T. (1975) 'Scaling loudness, noisiness and annoyance of aircraft noise', *Journal of Acoustical Society of America* 57: 930–934.

Berglund, B. and Lindvall, T. (1979) *Olfactory Evaluation of Indoor Air Quality*, Copenhagen: Danish Building Research Institute.

Bertin, J. (1979) 'Visual perception and cartographic transcription', *World Cartography* 15: 17–27.

Bianca, S. (1982) 'Traditional Muslim cities and Western planning ideology: an outline of structural conflict', in A. Serageldin and M. El-Sadek (eds) *The Arab City: Its Character and Islamic Cultural Heritage*, New York: Arab Urban Development Institute. 36–45.

Birchall, A. (1990) 'A whiff of happiness', *New Scientist*, 25 August: 44–47.

Blackhall, D. (1971) *The Way I See Things*, London: Baker.

Blades, M. and Spencer, C. (1986) 'Map use in the environment and educating children to use maps', *Environmental Education and Information* 5: 187–204.

Blades, M. and Spencer, C. (1988) 'How children find their way ...', *Environmental Education & Information* 7 (1): 1–14.

Blakemore, M. (1981) 'From wayfinding to map-making; the spatial information fields of aboriginal peoples', *Progress in Human Geography* 5 (1): 1–23.

Blauert, J. (1983) *Spatial Hearing: The Psychophysics of Human Sound Location* (trans. J. Allen), Cambridge: MIT Press.

Blaut, J. (1971) 'Space, structure and maps', *Tijdschrift voor Economische en Sociale Geografie* 64: 18–21.

Board, C. (1967) 'Maps as models', in R. Chorley and P. Haggett (eds) *Models in Geography*, London: Methuen, 671–725.

Bohm, D. (1983) *Wholeness and Implicate Order*, London: Ark.

Bombaugh, C.C. (1961) *Oddities and Curiosities*, New York: Dover Publications.

Borchett, J. (1987) 'Maps, geography and geographers', *Professional Geographer* 39 (4): 387–389.

Boring, E.G. (1942) *Sensation and Perception in the History of Experimental Psychology*, New York: Appleton-Century.

Bottomley, F. (1979) *Attitudes to the Body in Western Christendom*, Brighton: Harvester.

Bourdieu, P. (1977) 'The Berber house', in M. Douglas (ed.) *Rules and Meanings; An Anthropology of Everyday Knowledge*, Harmondsworth: Penguin, 98–110.

Brian, R. (1979) *The Decorated Body*, New York: Harper & Row.

Brookfield, H. (1969) 'On the environment as perceived', *Progress in Geography*, London: Edward Arnold.

Buttimer, A. (1993) *Geography and the Human Spirit*, Baltimore: The Johns Hopkins University Press.

Cain, W. (1978) 'History of research on smell', in C. Carterette and M. Friedman (eds) *Handbook of Perception: Volume 1*, New York: Academic Press, 197–229.

Cantacuzino, S. (1982) 'From courtyard to street: recent changes in the pattern of

Arab cities', in A. Serageldin and M. El-Sadek *The Arab City: Its Character and Islamic Cultural Heritage*, New York: Arab Urban Development Institute, 84–110.

Canter, D. and Lee, K.H. (1974) 'A non-reactive study of room usage in the modern Japanese apartment', in D. Canter and K.H. Lee (eds) *Psychology and the Built Environment*, London: Architectural Press.

Carpenter, E. (1973) *Eskimo Realities*, New York: Holt, Rinehart & Winston.

Chaney, D. (1990) 'Subtopia in Gateshead: the MetroCentre as a cultural form', *Theory, Culture & Society* 7: 49–68.

Chatwin, B. (1987) *The Songlines*, New York: Elisabeth Sifton Books/Viking.

Chesterton, G.K. (1958) *The Flying Inn*, London: Penguin.

Clarke, D.B. (1992) 'Reality, representation and simulation: a comment on theory and politics in Boyle and Hughes (1991)' *Area* 24 (2): 174–176.

Cohen, S., Glass, D. and Singer, J. (1973) 'Apparent noise, auditory discrimination and reading ability in children', *Journal of Experimental Social Psychology* 9: 407–422.

Cohen, S., Evans, G., Krantz, D., Stokols, D. and Kelly, S. (1981) 'Aircraft noise and children: longitudinal and cross-sectional evidence of adaptation to noise and the effectiveness of noise abatement', *Journal of Personality & Social Psychology* 40: 331–345.

Cooper, D. (1979) *Road to the Isles: Travellers in the Hebrides 1770–1914*, Glasgow: Richard Drew Publishing.

Cornish, V. (1928) 'Harmonies of scenery – an outline of aesthetic geography', *Geography* 14: 275–283, 382–394.

Cornish, V. (1935) *Scenery and the Sense of Sight*, Cambridge: Cambridge University Press.

Cosgrove, D. (1984) *Social Formation and Symbolic Landscape*, London: Croom Helm.

Cosgrove, D. (1985) 'Prospect, perspective and the evolution of the landscape idea', *Transactions of the Institute of British Geographers* 10: 45–62.

Cosgrove, D. and Daniels, S. (eds) (1989) *The Iconography of Landscape*, Cambridge: Cambridge University Press.

Crone, G. (1966) *Maps and their Makers: An Introduction to the History of Cartography*, London: Hutchinson.

Cuff, D. and Mattson, M. (1982) *Thematic Maps: Their Design and Production*, New York/London: Methuen.

Cullen, G. (1961) *Townscape*, London: Architectural Press.

Curry, M.R. (1991) 'Postmodernism, language, and the strains of modernism', *Annals of the Association of American Geographers* 81 (2): 210–228.

Daniels, S. (1989) 'Marxism, culture and the duplicity of landscape', in R. Peet and N. Thrift (eds) *New Models in Geography, Volume 2*, London: Unwin Hyman, 198–220.

D'Atri, D. (1975) 'Psychophysiological responses to crowding', *Environment and Behaviour* 7: 237–252.

Davis, W.M. (1909) *Geographical Essays*, Boston, MA: Little Brown.

Dear, M. (1988) 'The postmodern challenge: reconstructing human geography', *Transactions of the Institute of British Geographers* 13 (3): 262–293.

Debord, G. (1983) *Society of the Spectacle*, New York: Red & Black.

Deleuze, G. and Guattari, F. (1984) *Anti-Oedipus: Capitalism and Schizophrenia*, New York: Viking.

Derrida, J. (1982) *The Margins of Philosophy*, Harvester Press: Brighton.

Downs, R. and Stea, D. (1973) *Image and Environment*, Chicago: Aldine.

Durrell, L. (1992) *The Avignon Quintet*, London: Faber & Faber.

Eco, U. (1986) *Travels in Hyperreality, and Other Essays* (trans. W. Weaver) New York: Harcourt Brace Jovanovich.

Edgerton, S.Y. (1975) *The Renaissance Rediscovery of Linear Perspective*, London: Harper & Row.

Edstrom, K. (1990) *Saami Joik*, London: Radio 3 (broadcast 3 October).

Engen, T. (1972) 'Use of sense of smell in determining environmental quality', in W. Thomas (ed.) *Indicators of Environmental Quality*, New York: Plenum.

Engen, T. (1982) *The Perception of Odours*, New York/London: Academic Press.

Espeland, W. (1984) 'Blood and money: exploiting the embodied self', in J. Kotarba and A. Fontana (eds) *The Existential Self in Society*, Chicago/London: University of Chicago Press, 131–155.

Evans, G.W. and Howard, R.B. (1973) 'Personal Space', *Psychological Bulletin* 80: 334–344.

Evans, G.W. and Lepore, S. J. (1992) 'Conceptual and analytical issues in crowding research', *Journal of Environmental Psychology* 12: 163-173.

Featherstone, M. (ed) (1991) *The Body: Social Process and Cultural Theory*, London: Sage.

Fink, E. (1933) 'Die phenomenologische philosophie Husserls in der gegenwartigen Kritik Kantstudien', quoted in F.J. Wertz (1984) 'Procedures in phemenological research and the questions of validity' in C.M. Aanstoos (ed.) (1984) *Exploring the Lived World: Readings in Phenomenological Psychology*, West Georgia College: Studies in the Social Sciences XXIII, 29–48.

Folch-Serra, M. (1990) 'Place, voice, space: Mikhail Bakhtin's dialogical landscape', *Environment and Planning D: Society and Space* 8: 255–274.

Foucault, M. (1970) *The Order of Things: An Archaeology of the Human Sciences*, Andover, Hants: Tavistock Publications.

Foucault, M. (1972) *The Archaeology of Knowledge*, Andover, Hants: Tavistock Publications.

Foucault, M. (1973) *Discipline and Punishment: The Birth of the Prison*, New York: Vintage.

Foucault, M. (1975) *The Birth of the Clinic: An Archaeology of Medical Perception*, New York: Vintage/Random House.

Foucault, M. (1979) *The History of Sexuality* (3 vols), London: Allen Lane.

Foucault, M. (1980) *Power/Knowledge: Selected Interviews and Other Writings 1972–1977* (ed. C. Gordon), New York: Pantheon.

Foucault, M. (1967/1986) 'Of other spaces', *Diacritics* 16: 22–27.

Frank, A. (1991) 'For a sociology of the body: an analytical review', in M. Featherstone (ed) (1991) *The Body: Social Process and Cultural Theory*, London: Sage, 36–102.

Friedberg, A. (1993) *Window Stepping: Cinema and the Postmodern*, Los Angeles: University of California Press.

Fry, D. (1979) *The Physics of Speech*, Cambridge: Cambridge University Press.

Gardner, J. and Bartlett, P. (1992) *Sensors and Sensory Systems for an Electric Nose*, New York/London: Kluwer Academic.

Gibbons, D. (1986) 'The intimate sense of smell', *National Geographic*, September: 324–360.

Gibson, J. (1968) *The Senses Considered as Perceptual Systems*, London: George Allen & Unwin.

Gibson, J. (1974) *The Perception of the Visual World*, Westport, Connecticut: Greenwood Press.

Gilbert, A.N. and Wysocki, C.J. (1987) 'The smell survey results', *National*

Geographic, October: 514–525.

Giorgi, A. (1970) *Psychology as Human Science*, New York: Harper & Row.

Gold, J. (1980) *An Introduction to Behavioural Geography*, Oxford: Oxford University Press.

Gould, P. and White, R. (1974) *Mental Maps*, Harmondsworth: Penguin.

Grano, J. (1929) 'Reine Geographie: eine methodologische Studie beleuchtet mit Beispielen aus Finnland und Estland', *Acta Geographica* 2 (2): 191–195.

Green, K. (1982) 'Effects of aircraft noise on reading ability of school age children', *Archives of Environmental Health* 37: 24–31.

Greene, G. (1971a) *A Sort of Life*, New York: Simon & Schuster.

Greene, G. (1971b) *Journey Without Maps*, London: Penguin.

Guelke, L. (1976) 'Cartographic communication and geographic understanding', *The Canadian Cartographer* 13 (2): 107–122.

Hagino, G., Mochizuko, M. and Yamamoto, T. (1987) 'Environmental psychology in Japan', in D. Stokols and I. Altman (eds) *Handbook of Environmental Psychology*, New York: John Wiley and Sons.

Hahn, H. (1989) 'Disability and the reproduction of bodily images: the dynamic of human appearances', in J. Wolch and M. Dear (eds) *The Power of Geography: How Territory Shapes Social Life*, Boston: Unwin Hyman, 370–388.

Hall, E. (1969) *The Hidden Dimension*, London: Bodley Head.

Hargreaves, J. (1987) 'The body, sport and power relations', in J. Horne (ed.) *Sport, Leisure and Social Relations*, London: Routedge & Kegan Paul.

Harley, J.B. (1983) 'Meaning and ambiguity in Tudor geography', in S. Tyake (ed.) *Map Making 1500–1650*, London: British Library, 25–45.

Harley, J.B. (1989/1992) 'Deconstructing the map', in T.J. Barnes and J.S. Duncan (eds) *Writing Worlds: Discourse, Text and Metaphor in the Representation of Landscape*, London: Routledge, 231–247.

Hart, R. (1979) *Children's Experience of Place*, New York: Irvington.

Hart, R. and Moore, G. (eds) (1976) 'The development of spatial cognition: a review', in R. Downs and D. Stea (eds) *Image and Environment*, Chicago: Aldine, 246–88.

Harvey, D. (1989) *The Condition of Postmodernity: An Enquiry into the Origins of Cultural Change*, Oxford: Basil Blackwell.

Harvey, P.D.A. (1980) *The History of Topographical Maps: Symbols, Pictures and Surveys*, London: Thames & Hudson.

Haynes, R. (1981) *Geographical Images and Mental Maps*, London: Macmillan Education.

Heath, S. (1981) *Questions of Cinema: Communication and Culture*, London: Macmillan.

Heft, H. (1988) 'Gibson's ecological theory of perception', *Journal of Environmental Psychology* 8: 325–334.

Heidegger, M. (1966) *Discourse on Thinking*, New York: Harper Torchbooks (Harper & Row).

Heidegger, M. (1983) *Being and Time*, Southampton: Basil Blackwell.

Herbertson, A.J. (1905) 'The major natural regions', *Geographical Journal* 25: 300–310.

Herbertson, A.J. (1965) 'The higher units: a geographical essay', *Geography* 50: 332–42.

Herman, L. (1973) *British Landscape Painting in the Eighteenth Century*, London: Faber.

Herriot, J. (1977) *All Things Wise and Wonderful*, New York: St Martin's Press.

Hewison, R. (1989) *The Heritage Industry*, Methuen: London.

Higson, A. (1984) 'Space, place and spectacle', *Screen* 25: 2–21.

Hill, M.H. (1985) 'Bounded to the environment: towards a phenomenology of sightlessness', in D. Seamon and R. Mugerauer (eds) *Dwelling, Place and Environment*, Dordrecht: Martinus Nijhoff Publishers, 99–112.

Holt-Jensen, A. (1981) *Geography: Its History and Concepts*, London: Harper & Row.

Hopkins, J. (1990) 'West Edmonton Mall: landscape of myths and elsewhereness', *Canadian Geographer* 34 (1): 2–17.

Houston, J.M. (1982) 'Foreword', in A.J. Christopher, *South Africa*, London/New York: Longman, xii–xiii.

Hsu, Mei-Ling and Porter, P.W. (1971) 'Computer mapping and geographic cartography', *Annals of the Association of American Geographers* 61: 796–799.

Hull, J. (1990) *Touching the Rock: An Experience of Blindness*, London: SPCK Publishing.

Husserl, E. (1983) *Ideas Pertaining to Pure Phenomenology and to a Phenomenological Philosophy* (trans. F. Kersten) The Hague: Martinus Nijhoff.

Huxley, A. (1978) *Those Barren Leaves*, London: Granada.

Ihde, D. (1976) *Listening and Voice*, Athens, Ohio: Ohio University Press.

Irigaray, L. (1978) 'Interview with Luce Irigaray', in M.-F. Hans and G. Lapouge (eds) *Les Femmes, La Pornographie et L'Erotisme*, Paris.

Irigaray, L. (1985a) *The Spectrum of Other Women*, Ithaca, New York: Cornell University Press.

Irigaray, L. (1985b) *The Sex Which Is Not One*, Ithaca, New York: Cornell University Press.

Ittleson, W. (1974) *An Introduction to Environmental Psychology*, New York: Holt, Rinehart & Winston.

James, P. (1972) *All Possible Worlds*, Indianapolis: Bobbs-Merrill.

Jameson, F. (1984) 'Postmodernism, or the cultural logic of late capitalism', *New Left Review* 146: 53–92.

Jeans, D. (1974) 'Changing formulations of the man–environment relationship in Anglo-American geography', *Journal of Geography* 73 (3): 36–40.

Johnson, L. (1983) 'Bracketing lifeworlds: Husserlian phenomenology as geographical method', *Australian Geographical Studies* 21 (1): 101–108.

Johnson, N.B. (1989) 'The garden in Zuisen Temple, Japan', *Journal of Garden History* 10: 214–236.

Johnson, N.B. (1990) 'Geomancy, sacred geometry and the idea of a garden, Japan', *Journal of Garden History* 9: 1–19.

Johnston, R. (1983a) *Geography and Geographers: Anglo-American Human Geography since 1945* (2nd edn) London: Edward Arnold.

Johnston, R. (1983b) *Philosophy and Human Geography: An Introduction to Contemporary Approaches*, London: Edward Arnold.

Kariel, H. (1980) 'Mountaineers and the general public: a comparison of their evaluation of sounds in the recreational environment', *Leisure Studies* 3 (2): 155–167.

Kariel, H. (1990) 'Factors affecting response to noise in the outdoor recreational environment', *Canadian Geographer* 34 (2): 142–149.

Katsuki, Y. (1982) *Receptive Mechanisms of Sound in the Ear*, Cambridge: Cambridge University Press.

Kaufman, L. (1975) *Sight and Mind: An Introduction to Visual Perception*, New York: Oxford University Press.

Kern, S. (1974) 'Olfactory ontology and scented harmonies: on the history of smell', *Journal of Popular Culture*, Spring: 814–824.

Kikusawa, Y. (1979) 'Westernized home life style in Japan', *Home Economics*

Research Journal 7: 346–355.

Kirk, W. (1963) 'Problems of geography', *Geography* 48: 357–371, also in E. Jones (ed.) (1975) *Readings in Social Geography*, Oxford: Oxford University Press, 91–103.

Knapp, P.H. (1948) 'Emotional aspects of hearing loss', *Psychosomatic Medicine* 10: 203–222.

Korte, C. and Grant, R. (1980) 'Traffic noise, environmental awareness, and pedestrian behaviour', *Environment and Behaviour* 12: 408–420.

Kryter, K.D. (1985) *The Effects of Noise on Man* (2nd edn) London: Academic Press/ Harcourt Brace Jovanovich.

LAC (London Arts Council) (1986) *Looking into Paintings*, published to accompany the LAC exhibition of the same name.

Laing, R.D. (1964) *The Divided Self*, Baltimore: Tavistock Publications.

Lane, H. (1988) *When the Mind Hears: A History of the Deaf*, London: Penguin.

Lash, S. (1988) 'Discourse or figure? Postmodernism as a regime of significance', *Theory, Culture & Society* 5 (2–3): 311–336.

Lawless, H. and Engen, T. (1977) 'Associations to odors: interference, mnemonics and verbal labeling', *Journal of Experimental Psychology: Human Learning and Memory* 3 (1): 52–59.

Lefebvre, H. (1991) *The Production of Space* (trans. D. Nicholson-Smith) Oxford: Blackwell.

Lindvall, T. (1973) 'Sensory measurement of ambient traffic odours', *Journal of the Air Pollution Control Association* 23: 697–700

Long, S. (1992) 'The sense of sight', *National Geographic*, January: 283–310.

Long, R. and Cork, R. (1988) *Third Ear: Interview with Richard Long, by Richard Cork*, transcript, BBC Radio 3.

Lopez, B. (1986) *Arctic Dreams*, London: Macmillan.

Lowenstein, O. (1966) *The Senses*, London: Penguin.

Lowenthal, D. (1961) 'Geography, experience and imagination', *Annals of the Association of American Geographers* 51 (3): 241–260.

Lusseryan, J. (1963) *And There Was Light*, Boston: Little, Brown & Company.

Lusseryan, J. (1973) *The Blind in Society and Blindness: A New Seeing of the World*, New York: Proceedings 27, The Myrin Institute.

Lynch, K. (1960) *The Image of the City*, Cambridge, Mass.: MIT Press.

Lyotard, J.-F. (1971) *Discourse, Figure*, Paris: Klincksieck.

Lyson, K. (1978) *Your Hearing Loss and How to Cope With It*, New York: Fraeger.

Malancioiu, I. (1985) *Across the Forbidden Zone*, Bucharest: Editura Eminescu.

Maitland, B. (1985) *Shopping Malls: Planning and Design*, London: Construction Press.

Matthews, M.H. (1980) 'The mental maps of children', *Geography* 65: 165–179.

Matthews, M.H. (1984) 'Environmental cognition of young children', *Transactions of the Institute of British Geographers* 9: 89–106.

McLuhan, M. (1962) *The Gutenberg Galaxy*, Toronto: University of Toronto Press.

Meinig, D. (1979) *An Interpretation of Ordinary Landscapes*, Oxford: Oxford University Press.

Merleau-Ponty, M. (1962) *The Phenomenology of Perception* (trans. C. Smith) London: Routledge & Kegan Paul.

Mikesell, M.W. (1968) quoted by D. Cosgrove (1984) *Social Formation and Symbolic Landscape*, London: Croom Helm, 31.

Milne, L. and Milne, M. (1962) *The Senses of Animals and Men*, New York: Atheneum.

Mindel, E. and Vernon, M. (1987) *They Grow In Silence: Understanding Deaf Children and Adults* (2nd edn), Boston, Mass.: College Hill Press.

Mitchell, B. (1979) *Geography and Resource Analysis*, London/New York: Longman.

Monmonier, M.S. (1981) *Maps, Distortion and Meaning*, Washington, DC: Association of American Geographers.

Montagu, A. (1971) *Touching: The Human Significance of the Skin*, New York/London: Columbia University Press.

Montcrieff, R.W. (1966) *Odour Preferences*, New York: Wiley.

Moore, G. and Golledge, R. (eds) (1976) *Environmental Knowing: Theories, Research and Methods*, Stroudsburg, Penn.: Dowden, Hutchinson & Ross.

Murch, G.M. (1973) *Visual and Auditory Perception*, Indianapolis: The Bobbs-Merrill Co. Inc.

Murray, W.H. (1973) *The Islands of Western Scotland: The Inner and Outer Hebrides*, London: Longman.

Nash, P. (1986) 'The making of a humanistic geographer: a circuitous journey', in L. Guelke (ed.) *Geography and Humanistic Knowledge*, University of Waterloo, Dept. of Geography: Waterloo Lectures in Geography 2 (25): 1–22.

Norberg-Schultz, C. (1969) 'Meaning in architecture', in C. Jencks (ed.) *Meaning in Architecture*, London: Cresset Press.

Norberg-Schulz, C. (1980) *Genius Loci: Towards a Phenomenology of Architecture*, London: Academy Editions.

Northern, R. and Haskall, M. (1977) *Shopping Centres: A Developer's Guide to Planning and Design*, Reading: College of Estate Management.

O'Brien, M. (1989) *Reproducing the World*, London: Westview Press.

Ogawa, T. (1980) 'The geographical distribution and historical development of rural house types in Japan: a cultural geography', in The Association of Japanese Geographers (eds) *The Geography of Japan*, 161–183.

Ohlson, B. (1976) 'Sound fields and sonic landscapes in rural environments', *Fennia* 148: 33–45.

Omata, K. (1992) 'Spatial organisation of activities of Japanese families', *Journal of Environmental Psychology* 12: 259–267.

Ong, W.J. (1971) 'World as view and world as event', in P. Shepard, and D. McKinley (eds) *Environ/mental: Essays on the Planet as Home*, New York: Houghton, 61–79.

O'Neill, J. (1985) *Five Bodies*, Ithaca: Cornell University Press.

O'Rourke, P. (1990) 'Past imperfect', *Leisure News* 30: 12–14.

Peterson, N. (1972) 'Hunter-gatherer territoriality', *American Anthropologist* 77: 53-68.

Philo, C. (1992) 'Foucault's Geography', *Environment and Planning D: Society and Space* 10 (2): 137–162.

Piaget, J. and Inhelder, B. (1956) *The Child's Conception of Space*, New York: Humanities Press.

Pocock, D.C.D. (1981a) 'Sight and Knowledge', *Transactions of the Institute of British Geographers* 6: 385–393.

Pocock, D.C.D. (ed.) (1981b) *Humanistic Geography and Literature*, London: Croom Helm.

Pocock, D.C.D. (1983) 'Geographical fieldwork', *Geography* 68: 319–325.

Pocock, D.C.D. (1987) 'A sound portrait of a cathedral city', tape recording (presented to the Institute of British Geographers Annual Conference, 1987), Department of Geography, Durham University.

Pocock, D.C.D. (1988) 'The music of geography', in D. Pocock (ed.) *Humanistic Approaches to Geography*, Dept. of Geography, University of Durham: Occasional Paper (New Series) 22: 62–71.

Pocock, D.C.D. (1989) 'Sound and the geographer', *Geography* 74: 193–200.

Pocock, D.C.D. (1993) 'The senses in focus', *Area* 25 (1): 11–16.

Pollock, G. (1988) *Vision and Difference: Femininity and Feminism and the History of Art*, New York/London: Routledge.

Porteous, J.D. (1982) 'Approaches to environmental aesthetics', *Journal of Environmental Psychology* 12 (1): 53–82.

Porteous, J.D. (1985) 'Smellscape', *Progress in Human Geography* 9 (3): 356–378.

Porteous, J.D. (1986a) 'Bodyscape: the body–landscape metaphor', *Canadian Geographer* 30 (1): 2–12.

Porteous, J.D. (1986b) 'Intimate sensing', *Area* 18 (3): 250–251.

Porteous, J.D. (1990) *Landscapes of the Mind: Worlds of Sense and Metaphor*, Toronto: Toronto University Press.

Porteous, J.D. and Mastin, J.F. (1985) 'Soundscape', *Journal of Architectural Planning Research* 2: 169–186.

Poster, M. (1990) *The Mode of Information: Poststructuralism and Social Context*, Cambridge/Oxford: Polity Press/Basil Blackwell.

Punter, J.V. (1982) 'Landscape aesthetics: a synthesis and critique', in J.R. Gold and J. Burgess (eds) *Valued Environments*, London: George Allen & Unwin, 100–123.

Rayson, B. (1987) 'Emotional illness and the deaf' in V. Mindel and N. Vernon (eds) *They Grow in Silence* (2nd edn) Boston: College Hill Press, 65–102.

Rees, R. (1980) 'Historical links between geography and art', *Geographical Review* 70(1): 60–78.

Relph, E. (1976) *Place and Placelessness*, London: Pion.

Relph, E. (1981) *The Modern Urban Landscape*, London: Croom Helm.

Relph, E. (1982) 'Landscapes of consumer society', in A. Carlson and B. Sadler (eds) *Environmental Aesthetics: Essays in Interpretation*, Western Geographical Series 20, Dept. of Geography, University of British Columbia, Canada, 47–66.

Relph, E. (1985) 'Geographical experience and being-in-the-world: the phenomenological origins of geography', in D. Seamon and R. Mugerauer (eds) *Dwelling, Place and Environment*, Dordrecht: Martinus Nijhoff Publishers, 15–32.

Rice, S. (1991) *The Saxon Tapestry*, Sevenoaks, Kent: Hodder & Stoughton.

Robinson, A.H. and Pechenik, B.B. (1976) *The Nature of Maps: Essays Towards Understanding Maps and Mapping*, Chicago; Chicago University Press.

Rojek, C. (1993) 'Disney culture', *Leisure Studies* 12: 121–135.

Romey, W. (1987) 'The artist as geographer: Richard Long's Earth Art', *Professional Geographer* 39 (4): 450–456.

Rose, J. (1986) *Sexuality in the Field of Vision*, London: Verso Books.

Rosenthal, M. (1982) *British Landscape Painting*, Oxford: Phaidon.

Roth, I. and Froiby, J.P. (1986) *Perception and Representation: A Cognitive Approach*, Milton Keynes: Open University Press.

Rowles, G. (1976) *Prisoners of Space? Exploring the Geographical Experience of Older People*, Boulder, Colorado: Westview Press.

Ruark, R. (1964) *Uhuru*, London: Corgi Books.

Rundstrom, R.A. (1990) 'A cultural interpretation of Inuit map accuracy' *Geographical Review* 80 (2): 155–168.

Saarinen, T., Seamon, D. and Sell, J. (eds) (1984) *Environmental Perception and Behaviour: An Inventory and Prospect* (Dept of Geography, Research Paper 209), Chicago: Chicago University Press.

Samuels, M.S. (1978) 'Existentialism and human geography', in D. Ley and M. Samuels (eds) *Humanistic Geography: Problems and Prospects*, London: Croom Helm, 22–40.

Schafer, R.M. (1967) *Ear Cleaning*, Scarborough, Ontario: Berandol Music Ltd.

Schafer, R.M. (1977) *The Tuning of the World*, New York: Alfred A. Knopf.

Schafer, R.M. (1985) 'Acoustic space', in D. Seamon and R. Mugerauer (eds) *Dwelling, Place and Environment*, Dordrecht: Martinus Nijhoff Publishers, 87–98.

Schofer, J.P. (1975) 'Computer cartography and professional geographers', *The Professional Geographer* 27: 335–339.

Seamon, D. (1979) *A Geography of the Lifeworld*, London: Croom Helm.

Seamon, D. (1983) 'Doing phenomenology: possibilities and problems especially in relation to conventional positivist behavioural geography', paper presented to the national meeting of the Association of American Geographers, Denver, Colorado, April.

Seamon, D. and Mugerauer, R. (eds) (1985) *Dwelling, Place and Environment*, Dordrecht: Martinus Nijhoff Publishers.

Serageldin, A. and El Sadek, M. (1982) *The Arab City: Its Character and Islamic Cultural Heritage*, New York: Arab Urban Development Institute.

Shields, R. (1991) *Places on the Margin: Alternative Geographies of Modernity*, London: Routledge.

Skurnik, L.S. and George, F. (1967) *Psychology for Everyone*, Harmondsworth: Pelican/Penguin.

Soja, E.W. (1989) *Postmodern Geographies: The Reassertion of Space in Critical Social Theory*, London: Verso.

Sommer, R. (1969) *Personal Space: The Behavioural Basis of Design*, Englewood Cliffs, NJ: Prentice-Hall.

Southworth, M. (1969) 'The sonic environment of cities', *Environment and Behaviour* 1: 49–70.

Spencer, C. and Darvizeh, Z. (1983) 'Young children's place descriptions, maps and route-finding: a comparison of nursery school children in Iran and Britain', *International Journal of Early Childhood* 15: 26–31.

Stanley, G. (1988) *A Death in Tokyo*, Toronto: Bantam Books.

Stechow, W. (1968) *Dutch Landscape Painting of the Seventeenth Century* (2nd edn), London: Phaidon Press.

Stokols, D. (1976) 'The experience of crowding in primary and secondary environments', *Environment and Behaviour* 8: 49–86.

Stoddart, D. (1967) 'Organism and ecosystem as geographical models', in R. Chorley and P. Haggett (eds) *Models in Geography*, London: Methuen, 511–548.

Sullivan, T. and Gill, D. (1975) *If You Could See What I Hear*, New York: Harper & Row.

Taylor, D.R.F. (ed.) (1980) *The Computer in Contemporary Cartography*, New York: Wiley.

Tempest, W. (ed.) (1985) *The Noise Handbook*, London: Academic Press and Harcourt Brace Jovanovich Publishers.

Thomas, E. (ed.) (1962) *Selected Poems of Edward Thomas*, London: Faber & Faber.

Thrower, N. (1972) *Maps and Man*, Englewood Cliffs, NJ: Prentice-Hall.

Tolman, C. (1973) 'Cognitive maps in rats and man', in R. Downs and D. Stea (eds) *Image and Environment*, Chicago: Aldine, 27–50.

Tooley, R.V. (1978) *Maps and Mapmakers*, London: B.T. Batsford Ltd.

Tuan, Yi-Fu (1974) *Topophilia*, Englewood Cliffs, NJ: Prentice-Hall.

Tuan, Yi-Fu (1975) 'Images and mental maps', *Annals of the Association of American Geographers* 65: 205–213.

Tuan, Yi-Fu (1977) *Space and Place: The Perspective of Experience*, London: Edward Arnold.

Tuan, Yi-Fu (1979a) 'Sight and pictures', *Geographical Review* 69: 413–422.

Tuan, Yi-Fu (1979b) *Landscapes of Fear*, Minneapolis: University of Minneapolis Press.

Tuan, Yi-Fu (1993) *Passing Strange and Wonderful: Aesthetics, Nature and Culture*, Washington DC: Island Press/Shearwater Books.

Turner, B.S. (1984) *The Body and Society: Explorations in Social Theory*, Oxford: Basil Blackwell.

Unwin, K.I. (1975) 'Relation of observer and landscape in landscape evaluation', *Transactions of the Institute of British Geographers* 66: 131–134.

Von Horbestel, E.M. (1927) 'The unity of the senses', *Psyche* 7: 83–89.

Wakefield, N. (1990) *Postmodernism: The Twilight of the Real*, London: Pluto Press.

Walmsley, D. (1988) *Urban Living*, London: Longman Scientific and Technical.

Weissling, L.E. (1991) 'Inuit life in the Eastern Canadian Arctic, 1922–1942', *Canadian Geographer* 35 (1): 59–69.

Whitehead, A. (1938) *Modes of Thought*, New York: Macmillan.

Williams, R. (1965) *The Long Revolution*, Harmondsworth: Penguin.

Williams, R. (1973) *The City and Country*, London: Chatto & Windus.

Wohlwill, J.F. (1976) 'Environmental aesthetics: the environment as a source of effect', in I. Altman and J. Wohlwill (eds) *Human Behaviour and Environment* vol. 1, New York: Plenum Press.

Wolff, K.H. (1963) 'Surrender and aesthetic experience', *Review of Existential Psychology and Psychiatry* 3 (3): 209–226.

Wollen, P. (1980) 'Introduction: place in the cinema', *Framework* 13: 25.

Wood, D. (1993) *The Power of Maps*, London: Routledge.

Wood, L.J. (1970) 'Perception studies in geography', *Transactions of the Institute of British Geographers* 50: 129–142.

Wright, D. (1990) *Deafness: A Personal Account*, London: Faber (original 1969).

Wyburn, G.M., Pickford, R.W. and Hirst, R.J. (1964) *Human Senses and Perception*, Edinburgh: Oliver and Boyd.

Zimmerman, M.E. (1985) 'The role of spiritual discipline in learning to dwell on the earth', in D. Seamon and R. Mugerauer (eds) *Dwelling, Place and Environment*, Dordrecht: Martinus Nijhoff Publishers, 247–256.

INDEX

191